The Pottery from the Early Medieval Trading Site and Cemetery at Groß Strömkendorf, Lkr. Nordwestmecklenburg

RÖMISCH-GERMANISCHE KOMMISSION
DES DEUTSCHEN ARCHÄOLOGISCHEN INSTITUTS, FRANKFURT A. M.

Frühmittelalterliche Archäologie zwischen Ostsee und Mittelmeer

Band 1

Herausgegeben von

Sebastian Brather, Claus von Carnap-Bornheim,
Hauke Jöns, Christian Lübke, Friedrich Lüth,
Michael Müller-Wille & Karl-Heinz Willroth

in Verbindung mit Detlef Jantzen

RÖMISCH-GERMANISCHE KOMMISSION
DES DEUTSCHEN ARCHÄOLOGISCHEN INSTITUTS, FRANKFURT A. M.

Forschungen zu Groß Strömkendorf – III

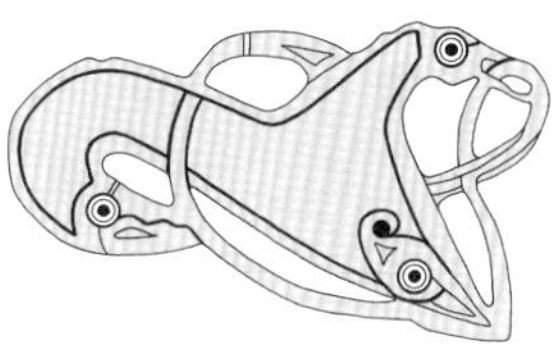

The Pottery from the Early Medieval Trading Site and Cemetery at Groß Strömkendorf, Lkr. Nordwestmecklenburg

by

Torbjörn Brorsson

Mit 51 Textabbildungen, 26 Tafeln und 23 Tabellen

Reichert Verlag Wiesbaden

Herausgeber: Sebastian Brather
Claus von Carnap-Bornheim
Hauke Jöns
Christian Lübke
Friedrich Lüth
Michael Müller-Wille
Karl-Heinz Willroth

Redaktion: Sonja Magnavita
Michèle Eller (techn. Red.)
Römisch-Germanische Kommission
Palmengartenstraße 10-12
60325 Frankfurt a. M.

Lektorat: Gabriele Franke
David G. Wigg-Wolf

Umschlagfoto: Sabine Suhr
Satz & Layout: Sonja Magnavita

Verlag: Reichert Verlag, Wiesbaden
Herstellung: Druckhaus "Thomas Müntzer", Bad Langensalza

Gedruckt mit Mitteln der Deutschen Forschungsgemeinschaft (DFG)

Bibliographische Information der Deutschen Bibliothek:
Die Deutsche Bibliothek verzeichnet diese Publikation in der Deutschen Nationalbibliographie; detaillierte bibliographische Daten sind im Internet über http://dnb.d-nb.de abrufbar.

ISBN 978-3-89500-751-4

Printed in Germany

Geleitwort zur Reihe

Der Raum zwischen Ostsee und Mittelmeer gehört zu den historisch hochinteressanten Gebieten Europas. Als traditionelles Kontaktgebiet zwischen Ost und West, Nord und Süd treffen hier seit jeher Menschen und Kulturen, Ideen und Güter zusammen. In nahezu allen Phasen der Geschichte war dies eine Zone des Austausches und der Kommunikation, deren Wege, Formen und Konjunkturen archäologisch vor allem durch die Verbreitung von – an den jeweiligen Fundorten – kulturell oder technologisch „fremden" Objekten zu erkennen ist. Die älteren Epochen des Raumes sind von besonderem Interesse, will man wesentliche Entwicklungen europäischer Geschichte dekodieren: ob es die Ausbreitung der Bandkeramik bis zur Odermündung und in das Pariser Becken während des frühen Neolithikums betrifft oder die Routen früher Metallhandwerker und die Vermittlung der Bronzegusstechnologie bis nach Skandinavien in der Bronzezeit, Fragen der Ostkontakte der Kelten in der Eisenzeit, die Auseinandersetzungen der Römer mit ihren „barbarischen" Nachbarn oder gleichzeitig bestehende Bernsteinrouten von der Ostsee bis nach Aquileia an der Adria, von denen Plinius der Ältere berichtete – stets war der Raum zwischen Ostsee und Mittelmeer von spezifischer Dynamik und Vermittlungsleistung.

Eine besondere Bedeutung kam diesem Gebiet während des frühen Mittelalters zu, als es zu einer Kontaktzone wurde, in der die westliche, karolingisch geprägte Welt auf slawische und reiternomadische Gesellschaften im Osten sowie skandinavische im Norden traf. Schriftliche Quellen zeichnen für diesen Teil Europas ein Bild ständiger Auseinandersetzungen, die zum einen durch das Streben nach Verteidigung, zum anderen durch Expansion vor allem des Karolingerreichs und seiner Nachfolger sowie anderer in dieser Zeit entstehender dynastischer Staaten ausgelöst bzw. bestimmt wurden. Die archäologischen Quellen stellen dabei ein entscheidendes Korrelat zur schriftlichen Überlieferung dar und vermögen aufgrund ihrer fachspezifischen Methodik Wesentliches auch zur interdisziplinären Diskussion beizutragen. In ihnen werden die unterschiedlichen, vor allem durch Handel und Austausch, aber auch durch Migration und Mobilität, Mission und Gegenreaktion ausgelösten Interaktionen zwischen Gemeinschaften und Siedlungsräumen deutlich. Eine von Alfried Wieczorek in Mannheim initiierte und von der Europäischen Union geförderte Ausstellung hat „Europas Mitte um 1000" publikumswirksam in den Mittelpunkt gestellt – jene Epoche früher Staatenbildung, die auf die „Slawisierung" Ostmitteleuropas folgte.

Die deutsche Forschung hat sich diesem Raum in besonderer Weise zugewandt, vor allem in jenen Gebieten, die unmittelbar östlich der Elbe zur DDR gehörten. Überregionale Forschungsergebnisse wurden in einer eigenen Schriftenreihe veröffentlicht. Infolge des politischen Umbruchs 1990 und der damit einhergehenden strukturellen Neuordnungen der Forschungslandschaft im vereinten Deutschland ist mit den „Schriften der Sektion für Vor- und Frühgeschichte" bzw. den „Schriften zur Ur- und Frühgeschichte" eine bewährte Publikationsreihe eingestellt worden, die zunächst von der *Sektion für Vor- und Frühgeschichte* der *Deutschen Akademie der Wissenschaften zu Berlin* und seit 1969 vom *Zentralinstitut für Alte Geschichte und Archäologie* (ZIAGA) der *Akademie der Wissenschaften der DDR* zwischen 1949 und 1994 in 46 Bänden herausgebracht wurde und beim Akademie-Verlag angesiedelt war. In diesen Bänden sind alle wichtigen, weit über das Gebiet der DDR hinausweisenden Forschungen vorgelegt worden.

Einen Teil der wissenschaftlichen Aufgaben des ZIAGA hat seit den frühen 1990er Jahren das *Deutsche Archäologische Institut* (DAI) übernommen. Räumlich und inhaltlich sind diese Aufgaben heute überwiegend auf das naturwissenschaftliche Referat an der Zentrale des DAI, auf die *Eurasien-Abteilung* sowie auf die *Römisch-Germanische Kommission* (RGK) des DAI übergegangen. Letztere beschäftigt sich satzungsgemäß innerhalb des DAI mit der Forschung zur Vor- und Frühgeschichte Alteuropas von den Anfängen bis in das Mittelalter.

Eine Reihe von Publikationsorganen des ZIAGA wurde aus rechtlichen Gründen eingestellt; es gelang nicht, einen neuen Vertragspartner zu finden, der sich mit dem die Rechte besitzenden Verlag zu einigen in der Lage sah. So liefen auch die „Schriften zur Ur- und Frühgeschichte" aus, zumal deren Publikationsanliegen in weiten Teilen Entsprechungen im nun gesamtdeutschen Fachbuchmarkt fanden. Aber nur in weiten Teilen: eine empfindliche Lücke entstand hinsichtlich der übergreifenden Forschungen zum Früh- und Hochmittelalter in den Gebieten östlich der Elbe, die die bestehenden westdeutschen Reihen nicht abdeckten. Diese Forschungen sind im herkömmlichen, fachlichen Sprachgebrauch als „Slawenforschungen" eingeführt. Die aus den fünf Forschungsstellen für Bodendenkmalpflege der DDR hervorgegangenen Landesämter für Archäologie oder für Bodendenkmalpflege in den fünf östlichen Ländern der

Bundesrepublik führten ihre Zeitschriften und Reihen weiter, und neue kamen hinzu, doch stehen dort auftragsgemäß regionale Interessen im Vordergrund.

Auch die RGK konnte diese Lücke mit ihren bestehenden Veröffentlichungsreihen nicht ohne weiteres schließen; so gibt sie zwar die „Römisch-Germanischen Forschungen" als Monographienreihe zur Vor- und Frühgeschichtsforschung heraus – doch hatte diese gut eingeführte Reihe nach 57 Bänden (heute 66) kein frühgeschichtliches Gepräge. Da eine Veränderung der Ausrichtung nicht wünschenswert erschien, blieb die Lücke für eine länderübergreifende Frühgeschichtsforschung bestehen. Eine auch nach Ostmitteleuropa ausgreifende Reihe, die Frühgeschichtsforschungen aufnehmen konnte, gab es nicht mehr. Die RGK entschied daher nach kurzer Beratung auf ihrer Jahressitzung 2008 einstimmig, diese empfindliche Lücke durch Gründung der Reihe „Frühmittelalterliche Archäologie zwischen Ostsee und Mittelmeer" zu schließen.

Wie wichtig dieses Vorhaben ist, zeigt das geplante Erscheinen gleich mehrerer Werke in kurzer Folge. Mit Forschungen zu Groß Strömkendorf, die von der Deutschen Forschungsgemeinschaft in den 1990er Jahren großzügig gefördert wurden, wofür auch an dieser Stelle allen Beteiligten nochmals gedankt sei, können gleich drei der ersten Bände dieser neuen Reihe gefüllt werden. Des Weiteren entstehen zurzeit Monographien zu Forschungen an anderen überregional wichtigen Plätzen, etwa zum befestigten Handelsplatz Parchim-Löddigsee (Dietlind Paddenberg), zur Burg Glienke (Sebastian Messal), zum von der Deutschen Forschungsgemeinschaft geförderten Projekt „Slawen an der unteren Mittelelbe", den Forschungen an der Burg Friedrichsruhe (Sebastian Messal) sowie zu einer Lutizenkonferenz (Christian Lübke). Weitere Bände sind in Vorbereitung, andere in Planung.

Das für die Reihe gewählte räumliche und zeitliche Programm wird im Titel deutlich; inhaltlich soll im Wesentlichen behandelt werden, was früher als Archäologie der Westslawen in der Forschung bezeichnet wurde. Es werden Monographien und Sammelwerke in deutscher und englischer Sprache erscheinen, wobei Zusammenfassungen in deutscher, englischer und russischer Sprache vorgesehen sind – individuelle und gut begründete Ausnahmen sind möglich. Die Herausgeber der neuen Reihe sind seit langem in der Forschung zur Frühgeschichte des Raumes zwischen Ostsee und Mittelmeer engagiert. Sie haben in mehreren Gesprächen diese Reihe konzipiert, werden über die Aufnahme von Werken entscheiden und tragen für die nächsten Jahre die Verantwortung für ihr Gedeihen.

Die RGK wird sich in Zukunft verstärkt um die Forschungen zur Frühgeschichte des Raumes zwischen Ostsee und Mittelmeer bemühen, Forschungen anregen sowie gemeinsame Forschungsprojekte mit in- und ausländischen Partnern entwickeln und durchführen. Dies hat die Kommission bei ihrer Jahrestagung 2008 beschlossen und der RGK den Auftrag erteilt, in den nächsten Jahren entsprechende Konzepte zu entwickeln. Mit der Reihe „Frühmittelalterliche Archäologie zwischen Ostsee und Mittelmeer" ist dafür gesorgt, dass eine Publikationsplattform all jenen offensteht, die sich mit diesem Raum und dieser Zeit beschäftigen.

Wir freuen uns auf die Zusammenarbeit mit dem Verlag Dr. Ludwig Reichert, danken Frau Dr. Sonja Magnavita für ihr engagiertes und beherztes Anpacken bei der Gesamtgestaltung der Reihe und der Redaktion des ersten Bandes und wünschen der neuen Reihe, Ihren Autorinnen und Autoren sowie den Leserinnen und Lesern eine gute Aufnahme.

Die Herausgeber der Reihe

Sebastian Brather
Claus von Carnap-Bornheim
Hauke Jöns
Christian Lübke
Friedrich Lüth
Michael Müller-Wille
Karl-Heinz Willroth

Für die Landesarchäologie Mecklenburg-Vorpommern als Mitherausgeber des vorliegenden Bandes
Detlef Jantzen

Contents

Vorwort der Projektleiter zum 3. Groß Strömkendorf-Band

Torbjörn Brorsson: The Pottery from the Early Medieval Trading Site and Cemetery at Groß Strömkendorf, Lkr. Nordwestmecklenburg

Mit der hier vorgelegten Untersuchung der Keramikfunde durch Torbjörn Brorsson wird die Veröffentlichung der Ergebnisse der Ausgrabungen und Forschungen im Bereich des frühmittelalterlichen Handelsplatzes Groß Stömkendorf an der Wismar Bucht – mit dem in den Fränkischen Reichsannalen genannten *emporium reric* wohl gleichzusetzen – fortgesetzt. Dieser Band ist – genauso wie Band 1 und 2, in denen die Tierknochenreste bzw. die Glasfunde diskutiert wurden – einer äußerst umfangreichen Materialgruppe gewidmet. Drei weitere Bände zu den Forschungen in Groß Strömkendorf befinden sich zurzeit in Vorbereitung. In Band 4 werden die von Astrid Tummuscheit analysierten Baubefunde und ihre Erkenntnisse zur strukturellen Entwicklung des Handelsplatzes vorgelegt. Band 5 ist dem nördlich des Handelsplatzes freigelegten Gräberfeld gewidmet und wird Manuskripte von Marcus Gerds (Archäologie) und Michael Wolf (Anthropologie/Paläopathologie) enthalten. In Band 6 der dem Fundplatz Groß Strömkendorf gewidmeten Reihe werden verschiedene naturwissenschaftliche und archäologische Beiträge unterschiedlichen Umfangs enthalten sein, die sich mit Fragen der topographischen und naturräumlichen Rahmenbedingungen, der Nahrungsmittelwirtschaft im Bereich des *emporium* und der Besiedlung in seinem unmittelbaren Umfeld beschäftigen werden. Darüber hinaus wird dieser Band auch Studien zu weiteren Materialgruppen, darunter zu Münzen und Gewichten, zu Bernsteinobjekten sowie Eisenschlacken und Resten der Buntmetallverarbeitung beinhalten. Die Veröffentlichung dieser Bände soll bis 2015 abgeschlossen sein.

Nachdem das Archäologische Landesmuseum mit Landesamt für Bodendenkmalpflege Mecklenburg-Vorpommern zu Beginn des Jahres 2006 umstrukturiert wurde und seine Eigenständigkeit verloren hatte, musste die dort angesiedelte wissenschaftliche Redaktion der „Beiträge zur Ur- und Frühgeschichte Mecklenburg-Vorpommerns", in denen Band 1 und 2 der Grabungsauswertungen von Groß Strömkendorf erschienen sind, zusätzliche Aufgaben übernehmen. Die zeitnahe Veröffentlichung der weiteren Manuskripte konnte nicht mehr sichergestellt werden. Deshalb wurde beschlossen, die Bände 3 bis 6 in die neu gegründete Schriftenreihe „Frühmittelalterliche Archäologie zwischen Ostsee und Mittelmeer" einzugliedern, deren erster Band mit dem vorliegenden Buch erscheint.

Bereits bei den Oberflächenabsammlungen und ersten Sondagen in Groß Strömkendorf in den 1980er und zu Beginn der 1990er Jahre wurden in großem Umfang Keramikfunde geborgen. Wie die Analyse dieses Materials durch Frank Wietrzichowski zeigte, beinhaltete es neben den mengenmäßig dominierenden frühslawischen Keramikwaren Sukow und Feldberg auch zahlreiche Keramikfragmente aus fränkischer und friesischer Produktion. Gemeinsam mit den ebenfalls vor Ort geborgenen, in skandinavischer und westeuropäischer Tradition stehenden Fibeln und Nadeln, wurden die Keramikfunde zu Recht als Beleg dafür gewertet, dass die im 8. und 9. Jh. an der Wismar Bucht ansässige Bevölkerung von Groß Strömkendorf über überregionale Kontakte verfügte.

Als es dann in der Zeit von 1995 bis 1999 Dank der Förderung durch die Deutsche Forschungsgemeinschaft möglich wurde, einen Teil der ursprünglich besiedelten Fläche großflächig freizulegen und ein benachbart gelegenes zeitgleiches Gräberfeld vollständig archäologisch zu untersuchen, vervielfachte sich auch der Umfang des vor allem aus Gräbern, Grabhügelumfassungsgräben, Häusern, Brunnen und Gruben stammenden keramischen Fundmaterials, das insgesamt 62300 Fragmente sowie einige, vor allem als Urnen genutzte vollständig rekonstruierbare Gefäße mit einem Gesamtgewicht von ca. 780 kg umfasst. Ein repräsentativer Teil der Scherben kann durch Vergesellschaftung mit typologisch bzw. numismatisch datierbaren Kleinfunden bzw. durch ihre Fundlage in dendrodatierten Befunden einem relativ engen Zeitfenster vom Ende des ersten Drittels des 8. bis zum Beginn des ersten Drittels des 9. Jh. zugeordnet werden. Damit entstammt es einem Zeitraum, in dem sich – nach gängiger Forschungsmeinung – die gesellschaftlichen Strukturen der im 7. Jh. in den südwestlichen Ostseeraum eingewanderten slawischen Siedlergruppen konsolidierten. Damals entstand das für die Frühphase der Urbanisierung des Nord- und Ostseeraums so bedeutsame Netz von küstennahen Handelsplätzen.

Wie die Analyse der Keramikfunde von vergleichbaren Fundplätzen zeigte – stellvertretend seien hier die Untersuchungen an den Keramikfunden von Haithabu (H. Steuer, K.-D. Hahn, W. Janssen), Ribe (S. Jensen und C. Feveile) und Emden (H. Stilke) genannt – haben Veränderungen im Kommunikationsverhalten bzw. in den Kontakten der jeweils ansässigen Bevölkerung auch im Massengut Keramik ihren Ausdruck gefunden, so dass im Fundmaterial aus Groß Strömkendorf ebenfalls ein hohes Erkenntnispotenzial zu vermuten war.

Vor diesem Hintergrund erscheint es logisch, dass sich die Projektleitung bereits während der laufenden Ausgrabungen bemühte, die Aufarbeitung der Keramikfunde zu organisieren und damit eine zeitnahe wissenschaftliche Bearbeitung zu ermöglichen. In den winterlichen Grabungspausen bildete deshalb die Säuberung, Inventarisierung, Restaurierung, zeichnerische Dokumentation und Erfassung des gesamten Materials in einer Datenbank eine der wichtigsten Aufgaben des Projektteams. Insbesondere Adelheid Lenius, Bettina Radtke, Astrid Wigger und Ilona Röpke waren mit diesen Arbeiten betraut; ihnen sind die auf den Tafeln abgebildeten Zeichnungen zu verdanken.

Darüber hinaus galt es für die wissenschaftliche Auswertung der Keramikfunde von Groß Strömkendorf einen Partner zu gewinnen, der nicht nur über die notwendige Erfahrung in der Untersuchung von entsprechenden Keramikkomplexen verfügte, sondern auch über die technischen Voraussetzungen, um die erforderlichen naturwissenschaftlichen Analysen durchzuführen. Dieser Partner wurde im renommierten Laboratory for Ceramic Research der Universität Lund (http://www.geol.lu.se/kfl/) unter Leitung von Anders Lindahl gefunden, das sich seit den 1970er Jahren auf die naturwissenschaftliche und kulturhistorische Analyse von Keramikfunden spezialisiert hat und das über erhebliche Erfahrung in der Untersuchung von Keramikensembles von Zentral- und Handelsplätzen des ersten nachchristlichen Jahrtausends aus unterschiedlichen Teilen des Ostseeraums verfügt.

Im Laboratory for Ceramic Research in Lund wurde Torbjörn Brorsson mit der Bearbeitung der Funde betraut; er war dafür bestens geeignet, da er sich bereits zuvor intensiv mit Keramikensembles anderer frühmittelalterlicher Siedlungen beschäftigt hatte. Nachdem eine anteilige Finanzierung der Untersuchungen durch das Land Mecklenburg-Vorpommern sichergestellt werden konnte, fasste er den Beschluss, die Untersuchungen auszuweiten und über das keramische Material eine Dissertation anzufertigen, in der er sowohl naturwissenschaftliche Analysemethoden anwenden als auch morphologisch-typologische Untersuchungen einsetzen wollte. Das Ergebnis dieser Arbeit wird mit diesem Band vorgelegt. Das von Anders Lindahl, Claus von Carnap-Bornheim und Michael Müller-Wille begutachtete Manuskript wurde im Sommersemester 2005 von der Mathematisch-Naturwissenschaftlichen Fakultät der Christian-Albrechts-Universität zu Kiel als Dissertation anerkannt und zum Druck freigegeben.

Wir danken vor allem Frau Sonja Magnavita für die rasche, hochprofessionelle Redaktion dieses Bandes. Ihrem Engagement ist es zu verdanken, dass in so kurzer Zeit die redaktionelle Fertigstellung des Werkes und seine Produktion erfolgt ist.

Der Deutschen Forschungsgemeinschaft danken wir für die Bereitstellung der Förderung nicht nur des Forschungsvorhabens Groß Strömkendorf, sondern auch der Publikationsmittel für diesen Band.

Hauke Jöns
Niedersächsisches Institut für historische Küstenforschung Wilhelmshaven

Friedrich Lüth
Römisch-Germanische Kommission des Deutschen Archäologischen Instituts/ Frankfurt a.M.

Michael Müller-Wille
Institut für Ur- und Frühgeschichte der Universität Kiel

Preface

Writing a doctoral thesis is surely a long trip. My trip started in 1997 when I first visited the site at Groß Strömkendorf. A year later, I started to document the pottery stored at Schloß Wiligrad at Lübstorf outside Schwerin. I have now reached the end of the road and owe several people my deepest gratitude.

The study of the pottery from Groß Strömkendorf is part of the project about this Early Medieval trading site. Prof. Dr. Drs. H. c. mult. Michael Müller-Wille at the University in Kiel and Priv.-Doz. Dr. Hauke Jöns at the Landesamt für Bodendenkmalpflege at Lübstorf have been in charge of the project. Prof. Müller-Wille has also been my supervisor and I am very grateful for all the help and important comments on my work. I am also much obliged to Priv.-Doz. Dr. Jöns who gave me the opportunity to work with the pottery and who has helped and supported me with lots of information for my thesis during all these years.

Ass. Prof. Anders Lindahl at the Laboratory for Ceramic Research at the University in Lund supported me during the years as a research student. He has been my supervisor in Lund and has commented on my work at different stages. There have been several discussions about pottery, ceramic craft and traditions.

I am also in deep gratitude to Ass. Prof. Birgitta Hulthén. She has encouraged me since I started working at the laboratory in 1995, and she has also commented and discussed parts of my thesis.

FD Dr. Ole Stilborg at the laboratory in Lund gave valuable remarks and inspiring criticism of the manuscript, for which I am very grateful.

Two persons that made the most of the thin sections are B.A. Emma Grönberg and M.Sc. Pia Sköld at the laboratory in Lund. B.A. Grönberg also helped me with some of the figures in my work. I would like to express my thanks to you both.

I am deeply indebted to Dr. rer. nat. Dietrich Meier at the University in Kiel/Mainz. He was the first person I met in Kiel in 1995. He was also the one who took me by car to Groß Strömkendorf one rainy and cold day in September 1997, telling me that we were going to a place I never had heard about, in all likelihood the former Reric.

FD Dr. Astrid Tummuscheit and M.A. Marcus Gerds at the University in Kiel gladly helped me with lots of archaeological and practical information about the settlement structure, the dendro-dates, the cemetery, the drawings, etc. I would like to express my deep gratitude to you both.

I also want to thank FD Dr. Torsten Kempke who checked the plates of the pottery. The comments and discussion have been very valuable for me and my work on the pottery.

I would like to thank FD Dr. W. J. H. Verwers at De Rijksdienst voor het Oudheidkundig Bodemonderzoek (ROB), State service Archaeological Investigations in Amersfoort in the Netherlands. I spent a day in Amersfoort and Dr. Verwers helped me with classification of the west-European pottery.

M.A. Christoph Keller at the Rheinisches Amt für Bodendenkmalpflege in Bonn examined several of the sherds classified as Vorgebirge-Eifel. The observations on their origin were of great importance.

Two persons that share my interest on Slavonic pottery are FD Dr. Mats Roslund at the department for Medieval Archaeology at the University in Lund and FD Dr. Blazej Stanisławski at the Archaeological department in Wolin. The discussions were always very interesting and inspiring.

I also would like to express my gratitude to FD Dr. Eva Andersson at the Archaeological department at the University in Lund. Dr. Andersson took time to examine some of my pottery sherds with textile imprints.

I am obliged to Museuminspector Claus Feveile at Den antikvariske samling in Ribe, Denmark. Claus organised my visit at the museum in November 2004, and he gave me important information about the Post Office excavation.

For several years I have been working with colleagues at the National Heritage Board in Lund and at Malmö Heritage in Malmö. I would like to thank you all (no one mentioned, no one forgotten). It has been a pleasure working with you.

I would like to mention B.A. Lars Jönsson and M.A. Joakim Thomasson at Malmö Heritage, M.A. Niklas Ytterberg at Bohuslän Museum in Uddevalla, FD Dr. Niklas Stenbäck at SAU in Uppsala and B.A. Maj-Lis Nilsson in Stockholm. Thank you for being friends and colleagues during the last years.

Anna Meier, Bernd Frose, Adelheid Lenius, Bettina Radtke, Ilona Röpke and Astrid Wigger have made all the excellent drawings of the ceramics from Groß Strömkendorf. They are an important part of the work and I thank you all.

Daniel Fugelsang examined the manuscript from a linguistic point of view. I owe you my thanks.

My parents, Ronny and Gertrud Brorsson, have closely followed my work during all the years. Special thanks for all your support, e.g. taken care of Emma and Elin while I needed time at the computer or in Kiel. Finally, the three girls in my life, Cecilia, Emma and Elin. You are the sunshine of my life and the ones that make me realise that there is more on earth than pottery sherds. My life and this work would not be the same without you!

Introduction

Groß Strömkendorf is situated approximately 7 km north of Wismar in Mecklenburg-Vorpommern, on the shores of the Baltic Sea. The topography of the area is more or less flat and the village lies close to Wismar Bay, in the shadow of the island of Poel *(Fig. 1)*. The area around the village conceals a very long history, containing archaeological finds from almost all prehistoric phases to the Early Medieval period.

Archaeological excavations during the 1980s and 1990s showed that, during the Early Medieval period, near today's village, there was a trading site with widespread connections in the Baltic area. The archaeological evidence clearly indicates that the trading site was settled during the 8th and early 9th century. During this period, the area was part of the territory of the Slavonic Obodrites, who have been in control of the whole western part of Mecklenburg. According to the find material, the site had been in close contact with Scandinavia, the Saxon and Frisian regions in today's lower Saxony and Schleswig-Holstein, as well as the Rhineland (Jöns 2000, 403). Several aspects of the archaeological artefacts and remains suggest that this site is identical with the *emporium reric*, mentioned in the Frankish annals of the early 9th century (id. 1998a, 131, with further references).

The finds and features discovered during the excavations show that the site was most likely part of some type of trading network. The importance of the site and its position in Northern Europe can be examined through studying different aspects of the excavations, e.g., the structures, metal finds, bones, glass, ceramics, botanical remains etc.

Detailed analyses and studies on the glass finds by A. Pöche (2005) and the animal bones by U. Schmölcke (2004) have already been published as monographs as the first two volumes of the Groß Strömkendorf-series in "Beiträge zur Ur- und Frühgeschichte Mecklenburg-Vorpommerns". Further analyses, i.e., on the settlement structure, the graveyard, the amber finds, on the geological and environmental development have already been completed and are going to be published in the present series, "Frühmittelalterliche Archäologie zwischen Ostsee und Mittelmeer". So, this study on the pottery from Groß Strömkendorf has been a part of an interdisciplinary research project.

Both the trading site and the cemetery at Groß Strömkendorf are dated to the 8th and early part of the 9th century. Hence, Groß Strömkendorf can be classified as an Early Medieval site. According to the Danish and Swedish chronological systems, the site was founded during the Vendel Period or the Germanic Iron Age and was abandoned during the Early Viking Age. The site of Groß Strömkendorf is, like Ribe and Birka, of great relevance for the frequent discussions about the dating of the beginning of the Viking Age (Feveile / Jensen 2000, 19; Ambrosiani / Clarke 1998; summing up Svanberg 2003).

Pottery as an archaeological source

Pottery is often regarded as the most common archaeological artefact category from excavations. It is often "classified" as a mass material, which means it takes time to scrutinise it. On the other hand, pottery as an archaeological artefact contains a lot of important information. Ceramic material can be used to create a picture of the society and their contacts, and of course to put it into a chronological framework. Pottery was most likely used every day in the household for cooking, storage and other purposes. It can also be considered as an item that appears in all social structures of a society. This is in contrast to other artefact categories, e.g., gold or silver. These finds may only reflect one part of the population or some special events.

Different pottery types with unique base and rim shapes were most likely well designed for their function. Vessels with round bases were perhaps made for cooking on a hearth, such as a hanging vessel or "Kugeltopf". Vessels with flat bases were perhaps intended for storing purposes. The same applies for the shape of the rim. Some types are better suited for cooking while others are better for storage.

In all likelihood, people were accustomed to the various functions related to different vessels. Moreover, the shape and decoration and their importance as identity markers are another aspect that has to be considered. Their significance when choosing one vessel over the other should not be underestimated. Is it the type of ware or the shape and decoration that determines which vessel to use? For instance, there are coil-built vessel types from

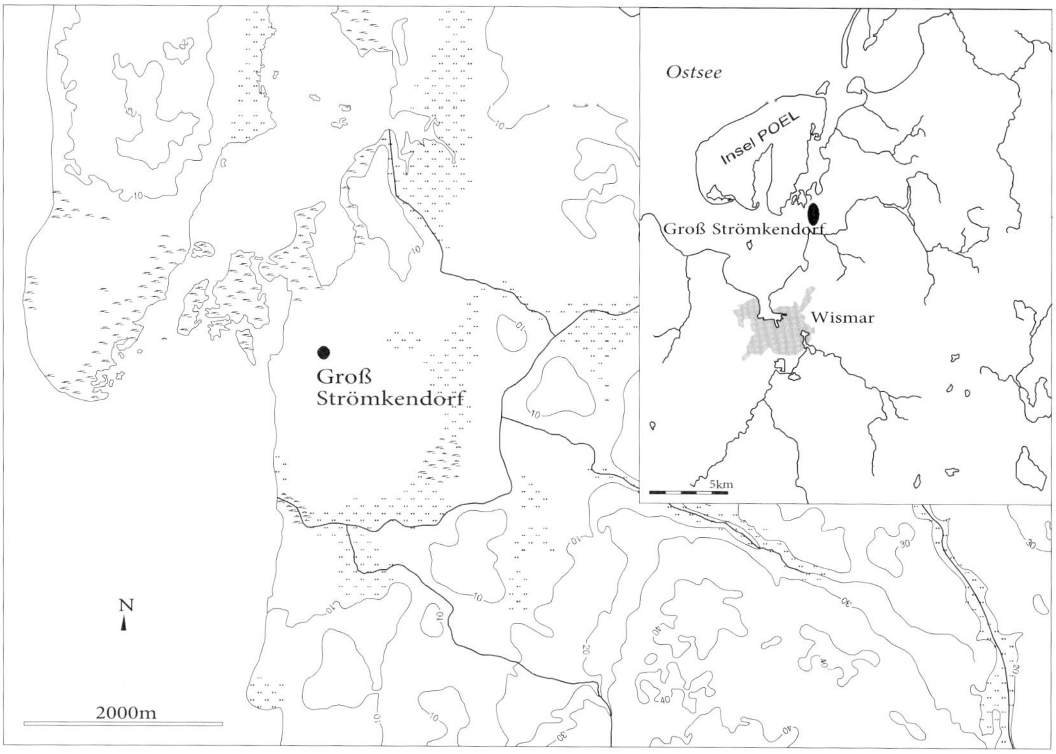

Fig. 1. Location of Groß Strömkendorf, Kr. Nordwestmecklenburg (Pöche 2005, Abb. 1).

Scandinavia as well as the Slavonic area that are of similar ware types (clay, temper in combination with firing technique). That means that the quality of the vessels is the same. Thus, from a practical point of view, they could easily be substituted by one another. Would then a Slavonic person use a Scandinavian vessel in her/his household, or is the tie to the cultural identity displayed in the shape and decoration too strong? The wheel-thrown vessels from Western Europe were of a new and to some extent better technology and, as such, seem to have been introduced as objects of status. The traditional handmade pots were probably better suited as cooking vessels, thus the imported vessels were likely used as containers or table ware for serving water, wine or beer. However, when an imported vessel is being introduced, this vessel is, in some aspects, most probably of a superior quality to the local ware.

The remains of wheel-thrown pottery are perhaps not the traces of people from Western Europe. The sherd material may instead indicate different trading routes within Europe, with the vessels produced in the Vorgebirge-Eifel area and transported to settlements, e.g., Dorestad in the Netherlands, and then further to other trading sites in Northern Europe.

Historical background and archaeological evidence at Groß Strömkendorf

Historical background

The Royal Carolingian Annals, written in Latin at the beginning of the 9th century, say:

"Year 808 AD. On Godfred's side stood during this campaign also the Slavonic people that is called the Wilzi, and in due of old hostility to their neighbours the Obodrites, they now peacefully had joined Godfred's armies: When he returned to his own kingdom, the Wilzi also went home together with the plundered goods they had taken from the Obodriten people. But before Godfred returned he destroyed a trading place located next to the sea, the trading place was by the Danes called Reric. It was a place that had revealed a great deal of taxes to his kingdom. He brought the trading people with him, let his ships set anchor and came together with his people to Sliasthorp" (after Skovgaard-Petersen 1981, 29). From the Frankish Annals of 809 we know in addition that the Obodrite duke Dražko was killed at *Reric*. After that event, *Reric* was not mentioned again in any contemporaneous written sources, so it likely was abandoned sometime at the beginning of the 9th century.

The location of *Reric* has been discussed by historians and archaeologists for a long time. First archaeological excavations, which aimed at locating *Reric,* were conducted in 1937 at a Slavonic hill fort at former Alt-Gaarz, about 14 km north of Groß Strömkendorf (summing up Jantzen / Schirren 1998, 71). These excavations were politically motivated because the national socialists wanted to change the town's name to *Reric*. Although the archaeological material from this excavation dates mainly from the 10th century – so it is extremely unlikely that Alt-Gaarz was the same place as *Reric* – Alt Gaarz changed its name to Rerik in 1938.

In addition, it has been suggested by different scholars that *Reric* could have been identical with the Slavonic hill forts of Mecklenburg or Starigard/Oldenburg, or that it was situated near Lübeck (summing up Barford 2001, 168) *(Fig. 2)*.

There are several written documents from the 8th and 9th century dealing with the political, economical and religious status of the different parts of Northern Europe. Most of these written sources are concerned with the expansion of the Carolingian Empire. However, there is also some information about the political system in the southern Baltic region and about the Slavonic and Scandinavian peoples living there. Most of these documents were written by people affiliated with the Carolingians or others and not by people living in the region.

Due to these sources, we know that the political map of Northern Europe was complex when *Reric* was abandoned at the beginning of the 9th century. The area where *Reric* most likely was situated bordered on three different cultural areas: the Slavonic, the Carolingian and the Scandinavian *(Fig. 3)*. The Carolingians, having ruled large parts of Western Europe for nearly 100 years, were then allied to a Slavonic tribe, the Obodrites. On the other side stood the Danish king Godfred, the Saxons, and the Slavonic tribe of the Wilzi. The conflicts in Northern Europe reached their climax during the reign of Charlemagne, between 768 and 814 (Brather 2001, 65).

All available information about the Baltic area with special focus on the Slavonic tribes has been compiled and interpreted by several archaeologists and historians. Some important studies are "Die Slawen in Deutschland" (Herrmann 1970), "The Early Slavs" (Barford 2001), "The Ancient Slavs: settlement & society" (Gojda 1991), "Archäologie der westlichen Slawen" (Brather 2001) and "Die Welt der Wikinger" (Sawyer / Sawyer 2002).

Scandinavian historical research had a special focus on Vikings travelling westward and eastward. Their conquests and trading ventures took place from the late 8th century to the 11th century. The Swedish Vikings ("Svear") from the Lake Mälaren region travelled eastward to present-day Russia and the Baltic countries. They followed the Russian rivers to the Byzantine Empire. Archaeological research and written sources suggest that the Vikings played an important role in the foundation of the Russian state (Oxenstierna 1998; Brorsson 2001). This eastern expansion resulted in the foundation of East-Slavonic dominions. Staraja Ladoga and Ruyrikovo Gorodishche/Novgorod, in present-day Russia, were among the most important centres. The expeditions to the east seem to have been more or less friendly and peaceful.

The Danish Vikings were of vital importance for the political structure in northwestern Europe. The raid on the monastery of Lindesfarne in Northumberland in 793 is the first documented Viking attack ever that was followed by a great expansion by Danish Vikings. In

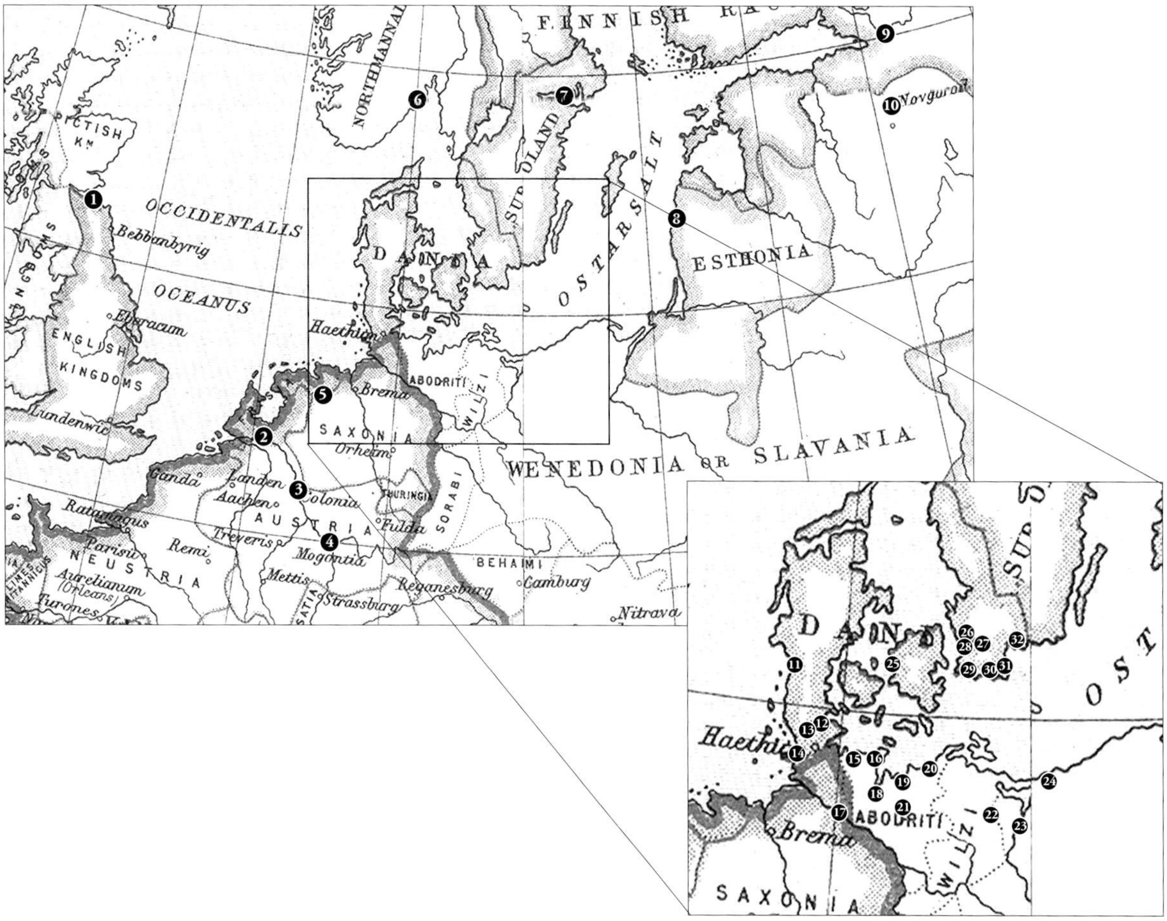

Fig. 2. Map of Northern Europe with the most important sites mentioned in the text.
1. Lindesfarne, 2. Dorestad, 3. Cologne, 4. Mainz, 5. Emden, 6. Kaupang, 7. Birka/Helgö, 8. Grobina. 9. Staraja Ladoga, 10. Ryurikovo Gorodishche, 11. Ribe, 12. Kosel, 13. Hedeby, 14. Elisenhof, 15. Scharstorf, 16. Starigard Oldenburg, 17. Hamburg, 18. Lübeck, 19. Groß Strömkendorf, 20. Rostock-Dierkow, 21. Mecklenburg, 22. Menzlin, 23. Szczecin, 24. Wolin, 25. Trelleborg (Slagelse), 26. Löddeköpinge, 27. Lund, 28. Bjärred, 29. Trelleborg (Scania), 30. Ystad, 31. Järrestad, 32. Åhus.

the following centuries, the Danes conquered and ruled a great part of the countries bordering the North Sea, including England. Norwegian Vikings colonised parts of Scotland, Ireland and northwestern England, Iceland and Greenland. The Viking Leif Eriksson reached as far as northern America.

According to the archaeological evidence, there can be no doubt that Scandinavians in these days also settled in the Slavonic areas, at the southern Baltic rim (summing up Jöns 2006; Jagodziński 2006). Actual research on sites such as Ralswiek on Rügen, Rostock Dierkow at the border of the Warnow river, Menzlin at the Peene river as well as Truso in the estuary of the Vistula and of course Groß Strömkendorf in the Wismar Bay, have given clear evidence of families of Scandinavian origin living there.

The history of the Slavonic people and culture is also one of expansion. The great Slav expansion took place much earlier than that of the Vikings, beginning already in the 6th century. It had its origin in the area north of the Carpathians in Eastern Europe. The Slavs moved to the south, as well as to the west and east. The expansion to the west stopped when they came in contact with the indigenous Germanic population and the Carolingian Empire. The river Elbe became an important frontier between the Slavonic population, the Saxons, and the Carolingians. During the late 7th century, the Slavonic expansion had reached as far as to the Baltic Sea coast, and for a long period the tribes were a potential military threat to the Carolingian Empire in the west. Some tribes within the West Slavonic community known from the written sources were the Obodrites, the Wilzi, the Ranes, the Wagrians, and the Pomeranians. The individual tribes had their own centres for administration and commercial purposes (Clarke / Ambrosiani 1993, 96; Müller-Wille 2002).

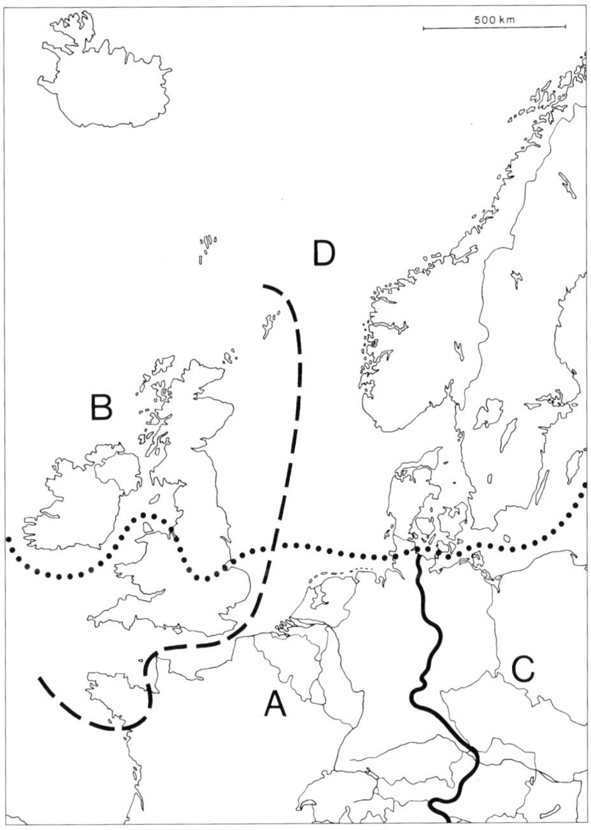

Fig. 3. Central, Western and Northern Europe. Sub-regions of Christianisation. A: Central and Western Europe; B: British Isles, Anglo-Saxon and Celtic (British, Irish, Pictish) domains; C: Eastern Central Europe, West Slavonic region; D: Northern Europe and Anglo-Scandinavian Britain (after Müller-Wille 1994, Fig. 2).

The involvement of different Slavonic tribes in the war between the Saxons and the Franks increased the Frankish interest in the Slavs. This was an important reason why the Franks in 789 invaded and conquered the Wilzi territory. During the 9th century, power centralised in the West Slavonic region under the strong influence of the Carolingian Empire (Barford 2001, 107).

Thanks to various written documents, of which the Royal Frankish Annals may be the most important, we know a lot about the contacts between the Carolingian Empire and the Slavs. However, there is only little information about the contacts within the West Slavonic tribes and about the contacts with northern Europe. The knowledge about the early Slavs' daily life, their burial traditions, social rankings etc. is even more limited (Brather 2004; Dulinicz 2006). The identity and ethnic origin of the Slavonic people is also an important issue.

The historical background of the political situation in northern Europe is not only important for the discussions on the possibility that the trading site of Groß Strömkendorf may be the site of the *emporium reric*. It also contains valuable information for the interpretation of the pottery material found. The pottery bears information about cultural origin, dating and further aspects, e.g., trade and contacts between different regions in Northern Europe. Combining this information with the historical sources may help to illustrate the political map in the region.

Archaeological evidence

In the middle of the 1950s, Schuldt published a study of the pottery from the Groß Strömkendorf site (Schuldt 1955), mainly concentrated on potsherds from the Funnel beaker culture and the Early Slavonic period in the region. The Sukow and Feldberg types dominated the Slavonic pottery inventory *(Fig. 4)*. Since the 1930s, the site was well known to amateur archaeologists who had collected a large amount of archaeological material, dominated by bones and Slavonic potsherds (Tummuscheit 2003b, 208). During the 1960s and 1970s, minor archaeological efforts were made at Groß Strömkendorf. At the end of the 1980s, Frank Wietrzichowski started to excavate in the area and soon realised that the place most likely was a trading site (Wietrzichowski 1993). Dendrochronological samples and the pottery suggested that the site belonged to the 8th century. In light of these finds, the dendrochronological datings, and the Frankish annals, Wietrzichowski interpreted the place as the former *Reric* (ibid. 1993, 43 ff.).

New excavations at Groß Strömkendorf, conducted by Hauke Jöns, began in 1995 and continued until 1999. They were organised by the "Institut für Ur- und Frühgeschichte der Universität Kiel" and the "Archäologisches Landesmuseum Mecklenburg-Vorpommern"; and funded by the "Deutsche Forschungsgemeinschaft" with the support of the state employment agency in Wismar (Jöns 1998b, 35). In connection with the archaeological excavations, geophysical investigations, analyses of aerial photographs as well as palaeobotanical, zoological and anthropological investigations were performed. A preliminary study on the new excavated ceramics from the second half of the 1990s was conducted by the author (Brorsson 2002b). The excavations were carried out each year between March and October or November (Tummuscheit 2003b, 11 ff.). The topsoil was removed by excavating machines until the virgin soil became visible. The discovered features were documented on a scale of 1:50 and stratigraphically excavated in levels of 0.10 m. The position of the finds was measured in x, y and z-coordinates. After washing, the finds were counted and divided into different artefact groups.

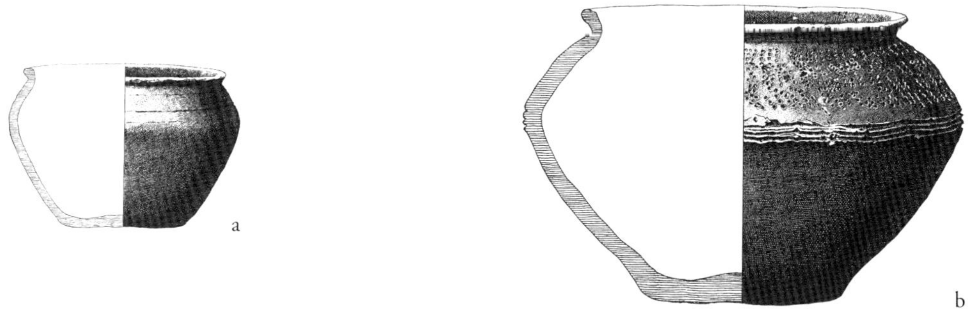

Fig. 4. Vessels of Sukow (a) and Feldberg (b) found at Groß Strömkendorf during excavations in 1955 (after SCHULDT *1955, Abb. 117, 119a). Scale 1:4.*

The outline and actual location of the trading place was confirmed early into the investigation. It was situated in a south-north direction along the shoreline *(Fig. 5)*. The main excavated area was a narrow and long band, approximately 25 m wide and 550 m long, parallel to the coastline (excavation area No. 3). This area most likely constituted the core of the trading settlement. Another area (excavation area No. 17), approximately 50 m north-east of the first trench, comprised the cemetery and the oldest phase of the trading activity.

A considerable part of the trading site has been destroyed by sea-level fluctuations and by the effects of coastal erosion. Only about 10 % of the former trading site have been exposed so far. Ca. 40 % have to be regarded as destroyed by coastal erosion.

The excavated area consists of more than 1,000 features, comprising pit houses, wells, hearths, ditches, and different types of pits. In some areas, cultural layers were also excavated. The total number of pit houses is about 100, and there are ca. 85 constructions interpreted as wells. Approximately 30 of these wells contained water-logged wood (TUMMUSCHEIT 2003b, 213).

The finds from the settlement are mostly of Slavonic origin, but a significant amount of the finds were imports. Apart from the pottery, lava querns from the Eifel region, sherds of funnel beakers made of glass, metal dress-fittings and weapons were also found, pointing towards a Carolingian and Saxon-Frisian origin. A large amount of objects came from Scandinavia. Some examples are whetstones from Norway or western Sweden, metal dress-fittings and a horse-shaped fibula dated to ca. 700 (KLEINGÄRTNER / JÖNS 2000). Finds of dirhams point to an Arabic contact. Other coins at Groß Strömkendorf are Frisian sceattas and Carolingian deniers.

Some of the identified crafts at Groß Strömkendorf are antler and bone working, most likely for the manufacture of combs. Glass working and objects of amber indicate that bead production took place at the settlement. Textile craft is represented by loom weights and spindle whorls. Metal work is attested by different types of slag, moulds and crucibles.

The pit houses and the wells are of great importance for the study of the pottery, since these types of constructions contain a large amount of ceramic finds and several wells are dated by dendrochronology.

Altogether, 95 samples of beech, oak, ash, pine and alder wood from 28 wells and one post have been dated by dendrochronology. The features are located in different areas of the excavation site, and the dates stretch mainly from the first half of the 8th century to 811 *(App. 1)*. A few wells could only be dated to after 654 or 722 due to the fact that the tree bark is missing and thus the exact date can not be established. The majority of the structures date between 761 and 786. These datings are certainly an important contribution to the interpretation of the Groß Strömkendorf site, the features and the artefacts, and in this special case, the distribution of the pottery. However, the dating of the wood does not always reflect the year the feature was built and the pottery was deposited. The dendrochronological method dates the year when the tree was cut, but the timber could have been used a considerable number of years later. In addition, the pottery was most likely only deposited after the wells were abondoned (TUMMUSCHEIT 2003a). It is also likely that ceramics from structures, e.g., wells from 780 and 806, may have been deposited during the same year, since timber from different years could have been used in the same well. This is the result of the archaeological interpretation of the settlement structures.

The structure of the trading site has been analysed by Astrid TUMMUSCHEIT (2003a). In the study, Tummuscheit divided the settlement into three different

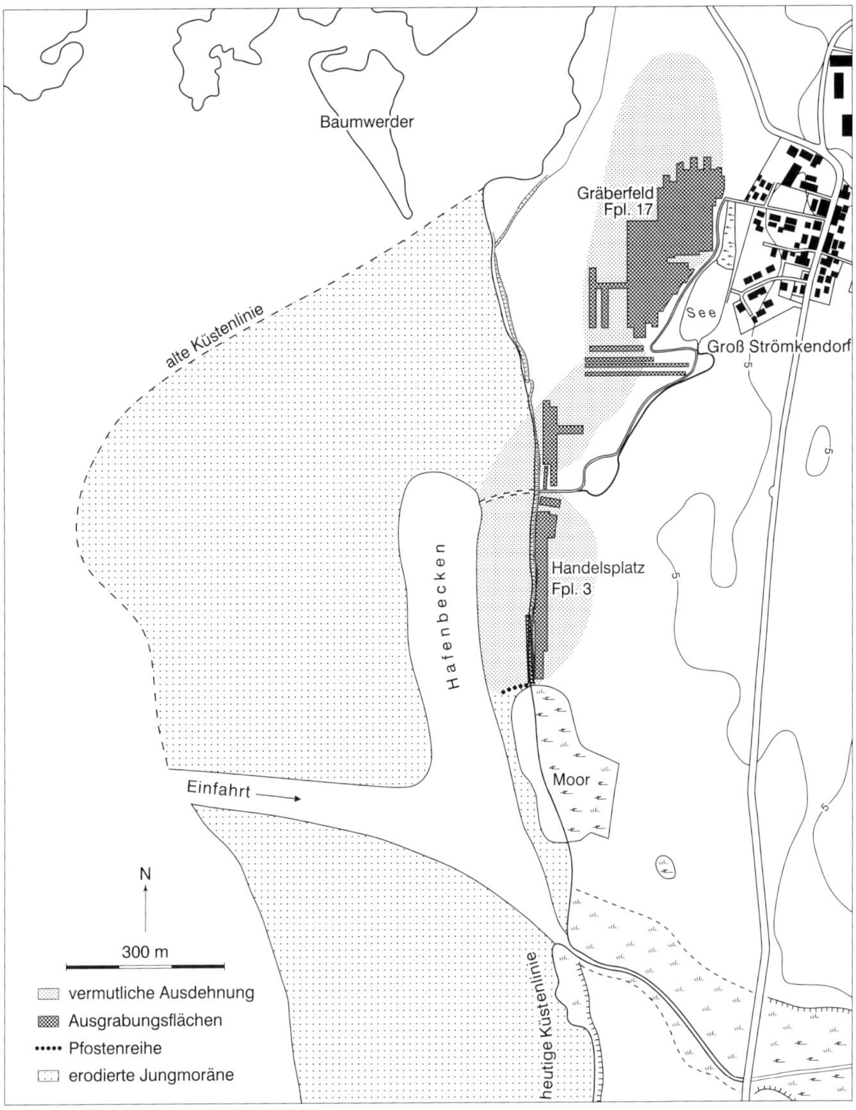

Fig. 5. The position of the trading site and cemetery at Groß Strömkendorf (after Jöns *2000).*

phases. Each phase contained pit houses and wells. The construction of the pit houses was studied in detail, thus leading to a further division based on the number of posts, hearths, and the position of the structures. The oldest settlement phase was dated to the first half of the 8th century. It was located in the north-eastern part, in area No. 17, the same area where the cemetery was discovered *(Fig. 6)*. The second phase, in the north of area No. 3, was dated to around 760. The last excavated phase was the southernmost part, dated from 780 until the end of the settlement at the beginning of the 9th century. According to the study of the settlement phases, the dating of the central area is uncertain. However, the dendrochronological datings and the study of the ceramics mainly indicate a quite late date.

A cemetery comprising more than 250 graves, dated from around the middle of the 8th century to the beginning of the 9th century, was exposed in the vicinity of the settlement. The cemetery represents a great variety of burial customs, with different types of cremation and inhumation graves (Paddenberg 2000; Gerds in prep.). There are inhumation graves with and without vessels. A few graves contain wooden coffins. One possible chamber grave indicating a West European origin was found. The cremation graves were divided into urn graves, cremation pits, cremation layers, and cremation deposits. The urn graves contained cremated bones, and, on some occasions, human bones were also found outside the urn. Many of the urns were completely destroyed, making it sometimes impossible to determine the urn type. The cremation pits contained ashes, cremated human bones and in some cases grave goods. The cremation layers were layers with ashes and cremated human bones. The cremation deposits are classified as concentrations of cremated human bones without a certain grave pit. They were mostly deposed into open ditches of mounds,

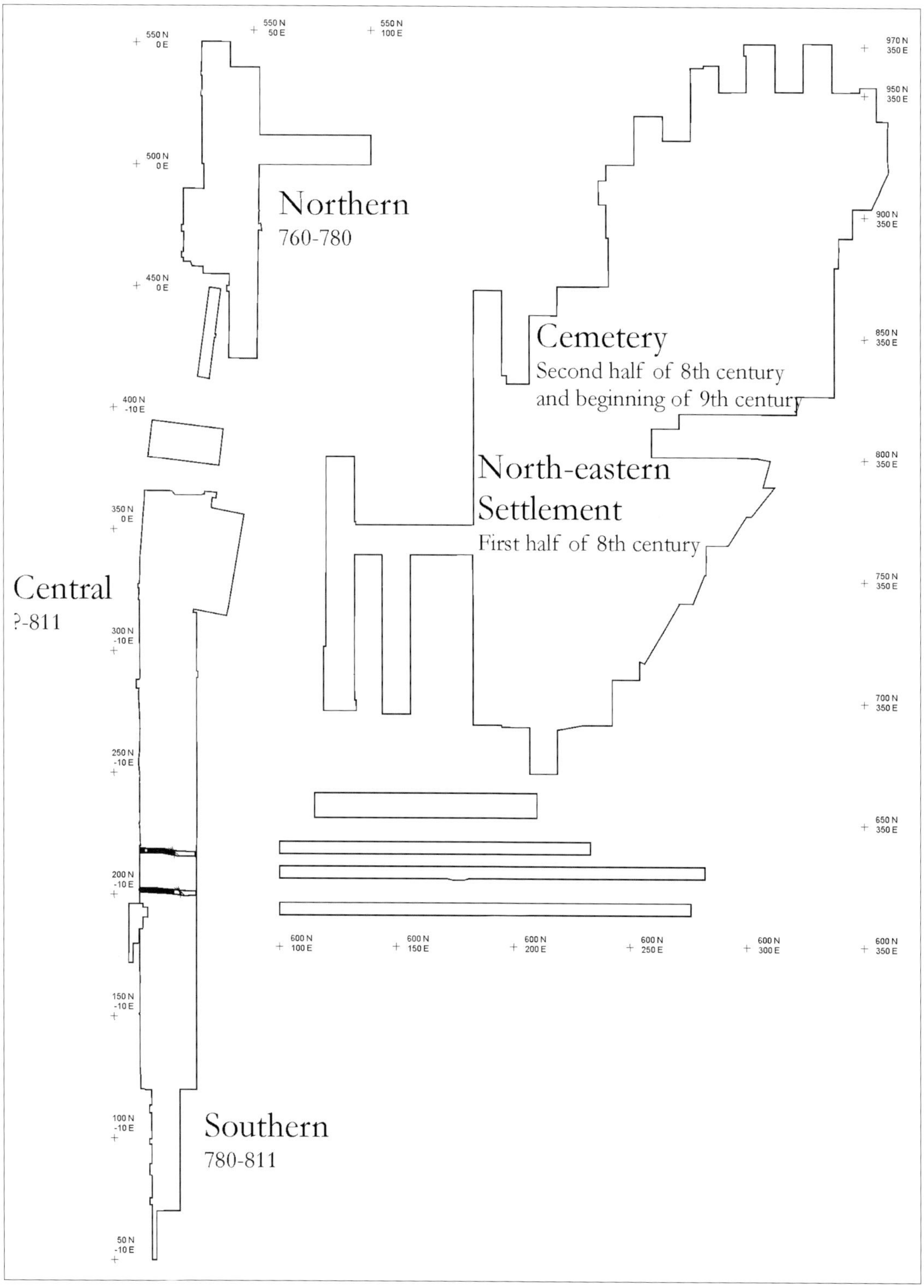

Fig. 6. *Preliminary datings of the settlement phases and the cemetery at Groß Strömkendorf. The cemetery and north-eastern area is situated northeast of the northern area. See Fig. 5.*

without any remains of a vessel or other grave goods. In addition, six boat graves and several animal graves were found. Some of the animal graves may be interpreted as sacrifices instead of graves (Jöns 1998b, 43). The boat graves are most likely connected to the Scandinavian population, since they were typical clinker-built Scandinavian vessels with iron rivets joining the overlapping planks (id. 1998a, 134). Some of the graves contained objects from different regions in Europe. The grave goods and the burial customs point towards contacts with the Frankish kingdom as well as with the Frisian, Scandinavian and Saxon settlement areas.

The cemetery is a significant contribution to the knowledge about the composition of the population of Groß Strömkendorf, being an important example of early urbanisation in the southern Baltic region. It is noteworthy that our information about the Slavs and their material culture as well as the hill forts is fairly good, whereas the knowledge concerning their rural settlements and burial customs is poor (Barford 2001, 199 ff.).

In 1999, aerial photographs of some kind of underwater cropmarks, in a distance of about 80 m from the seashore, were revealed. The filling of a roughly rectangular depression, 750 m long and 150 m wide, had enhanced the growth of water plants, showing as a dark anomaly on the photographs. This structure may be interpreted as the harbour basin of the settlement (Tummuscheit 2003b, 212).

Aims and problems

The pottery from Groß Strömkendorf is essential for interpreting contacts in Northern Europe during the Early Medieval time. Traditional studies have put a focus on the shape of the vessels and the decoration, and these elements should also be integrated in the study of the ware.

But other aspects of the pottery, e.g. ware, morphology, manufacturing technique (including the choice of raw material) and vessel building technique allow a look beyong dating and cultural affiliation. For example, the clay and temper material may indicate whether different vessels were made in the same area. Still, such analyses must be combined with a typological and/or chronological study. In general, this approach to ceramic material is more and more accepted (Daim 2000).

Archaeology is a research subject where the scientists reconstruct history by combining small pieces of information from various sources. The same holds true for the pottery. We normally do not have the opportunity to investigate complete vessels. It is instead the small pieces of a vessel, the sherds that are our base material, which we must investigate and interpret. In this study, the ware is of the same significance as the vessel shape and decoration.

It is important to record the clearest attributes of the vessel, such as decoration or rim shape. In addition, the ware can be analysed, especially in the comparion of ware types from the Baltic region to those from Western Europe. However, since people travelled and brought along skills and/or objects, ceramics found may be copies from other regions, making the reconstruction of history rather difficult. The same source critical aspects apply for other analyses. Clays from, e.g., northern Germany and Poland, are geologically similar to those from Scandinavia, because the ice during the last Ice Age caused a mixture of the soil in the Baltic area. This has an effect on how to analyse pottery sherds. Since exact methods of pinpointing the origin of the vessels are currently missing, the analyses must be compared with other pottery studies. The investigation of the pottery material from Groß Strömkendorf includes a quantitative study. However, there are also some problems to this aspect. The deposition in the pits, the houses, the wells and even in the graves during the Viking Age may not have been deliberate. Additionally, a pit may contain unusual pottery types, for which it could be difficult to establish whether they are from different period or represent some kind of special function. Other activities, such as modern agriculture or the changing of the sea level, could also affect the study.

These are some of the basic conditions when dealing with archaeological material and especially pottery. However, the artefacts and features must be analysed and investigated to gain a trustworthy picture of the settlement, the society and the people that stand behind the archaeological material. In consideration of this background, the following topics will be studied:

1. The settlement structure is a vital problem. Can the pottery investigation add to the understanding of this problem?

2. Are the traders among those buried at the cemetery? If so, can an investigation of the grave pottery shed some light on this aspect?

3. What are the differences and similarities between the cemetery vessels and the trading site pottery?

4. To what extent are the ceramics at Groß Strömkendorf of local or foreign origin?

5. What kinds of Slavonic pottery were found at Groß Strömkendorf, and what are the similarities and differences between the different types?

6. Can the pottery contribute to the discussion concerning Groß Strömkendorf being ancient *Reric*, as mentioned in ancient written sources?

Methods

The ceramics from Groß Strömkendorf have been scrutinised in order to extract as much information as possible. The above-mentioned questions will be answered by working with detailed documentations, different analysis methods and other information about the pottery craft and the society behind it, e.g., ethnoarchaeological sources. The documentation will mainly put a focus on the archaeological interpretation of the origin of the vessels. It will also be used to discuss the pottery craft during the 8th and early 9th century. The analyses will hopefully provide further details about the origin and the ceramic craft.

Combining different methods will deliver more adequate information. An important part is the statistical analysis. More than 62,300 sherds were recovered, and several parameters of the ceramics were measured, used for comparison with context information and the results of other analyses.

Documentation

All pottery found in Groß Strömkendorf during the large-scale excavations from 1995 to 1999 was handled in the same way. The sherds from each section were divided into pottery type and separated into rim, body and base sherds. In order to extract as much information as possible from the ceramic material, several parameters were recorded, based on the methods developed by B. Hulthén (1974). Weight, amount, sherd thickness, temper (type, amount, largest grain), vessel-building technique, shape of sherds, vessel type, part of vessel and, where applicable, decoration were recorded, as well as the original diameter of rims and bases and the vessel height. They were then entered in the computer programme dBase, followed by a more thorough input by the author using MS Access.

The classification of the Slavonic pottery is based on Schuldt's classification system (1956) with its main groups Sukow, Feldberg, Menkendorf and Fresendorf.

The rims were documented according to the system presented by T. Kempke (1981a, 37 Abb. 5) on the Slavonic material from Starigard/Oldenburg *(Fig. 7)*. By using a combination of Schuldt's and Kempke's classification systems, the Groß Strömkendorf material can easily be related to other studied materials. Rims that do not fit into Kempke's classification system were drawn and given a new classification number. The imported types were named either according to their presumed production area in Europe or according to the ware type.

A special study of the base sherds was conducted based on the shape of the base and the shape of the transition between base and body. The underside of the base sherds was also studied. In some cases, sand or textile impressions were identified.

The decoration was documented according to a system developed by Dirk Meier (1990, Abb. 19) on sherd material from Scharstorf in Schleswig-Holstein *(Fig. 8)*. Decoration types that do not fall within Meier's system were drawn and given new classification numbers. The location of the decoration was also recorded.

Casts were taken of different types of stamped impressions, made of a material normally used by dentists. It is a two-component substance called Provil Novo.

Statistical distribution

The evaluation of ceramic material contains an essential statistical part, based on the number and weight of the sherds. For this, it makes a difference whether sherds are thick walled, e.g., from cooking or storage vessels, or thin walled as from tableware. Based on the weight, the thicker sherd of the same size would have a higher share in the statistics. The same applies for the amount of sherds, since a larger vessel could produce more sherd material. Such misleading results need to be avoided.

This problem has been discussed by different scholars, and several systems for the quantification of pottery exist (Orton et al. 1993, 21 f.) A common method is to use vessel units to calculate the minimum number of vessels from a single context. This method has not been used on the pottery from Groß Strömkendorf, the main reason being the large amount of anonymous body sherds of either Slavonic or Scandinavian origin. Because of their similarity, it has been almost impossible to determine how many vessels the sherds represent. Instead, this inventory requires the use of other formulas to obtain a more precise picture of the ceramics (Lindahl 1986, 77 f.). The average sherd thickness is calculated with the formula in *Appendix 2*. As shown in *Appendix 3*, an adjusted weight is calculated to avoid that thicker sherds

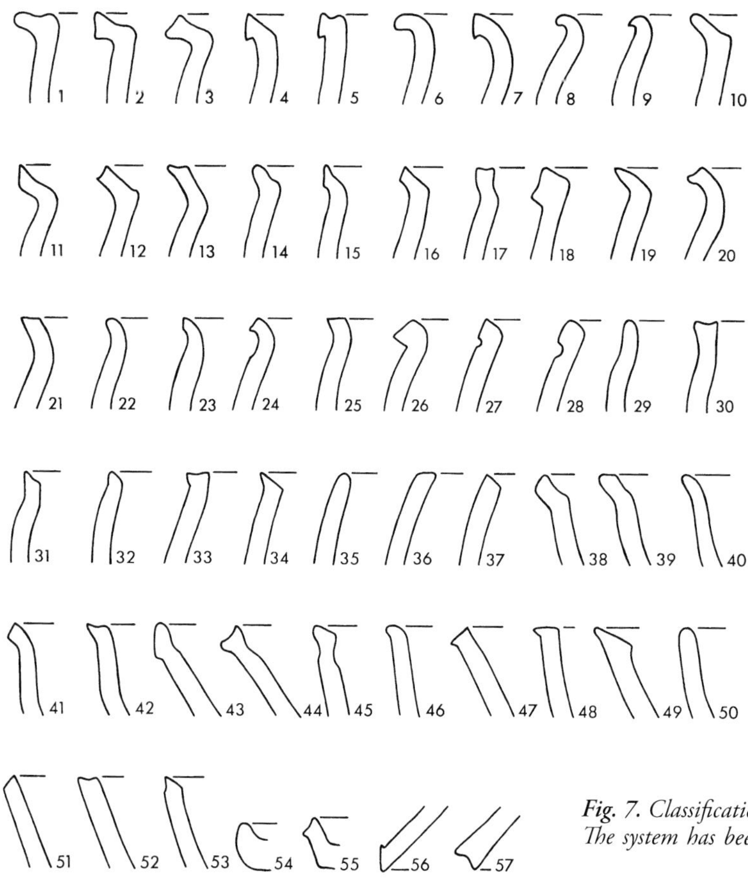

*Fig. 7. Classification system made on Slavonic ceramics at Oldenburg. The system has been used on the Slavonic, Scandinavian and Saxon pottery at Groß Strömkendorf (*Kempke *1981a).*

get a larger proportion in the statistical distribution than thin-walled sherds. As a standard reference thickness, 10 mm were chosen.

The fragmentation of the pottery from the features at Groß Strömkendorf was calculated. It can be used in mapping the intensity of the activity in any given area. In addition, it can give an indication about the use of a vessel type, a special vessel size or vessels of different sherd thickness, decoration, etc. The formula for calculating the mean sherd size is presented in *Appendix 4*.

Microscopy

A total of 177 thin sections of potsherds from Groß Strömkendorf were studied under a polarising microscope. According to the large Slavonic material most of the analyses were made on this material. Several analyses were also performed on Scandinavian pottery and on the other imported vessel types. The selection of analyses was also made on ceramics from the cemetery as well as from the dwelling site. Pottery from different periods was also choosen. The microscopy was made at magnifications between 25 xs and 630 xs, in parallel as well as in polarised light. The analyses were carried out by the author at the Laboratory for Ceramic Research, Institute of Geology at Lund University in Sweden. A mineralogical examination of a sample of sherds was conducted by D. Larsson at the Institute of Geology, Department of Mineralogy, at Lund University.

Microscopy under polarising microscope

Thin sections were made from each one of the selected sherds. For comparison reasons, the standard thickness of exactly 0.03 mm was applied. Under a polarising microscope it is possible to identify minerals in the silt and sand fractions of the sherds. But microscopy offers not only a petrographic study, it also shows other aspects of the pottery craft. It is possible to study different ware structures, such as vessel forming technique, clay type, difference between added and natural temper, as well as amount and grain size of the temper, organic matter e.g. diatoms and plant material, etc. All this information was recorded with the assumption that sherds of the same type of clay and temper were most likely from the same production area.

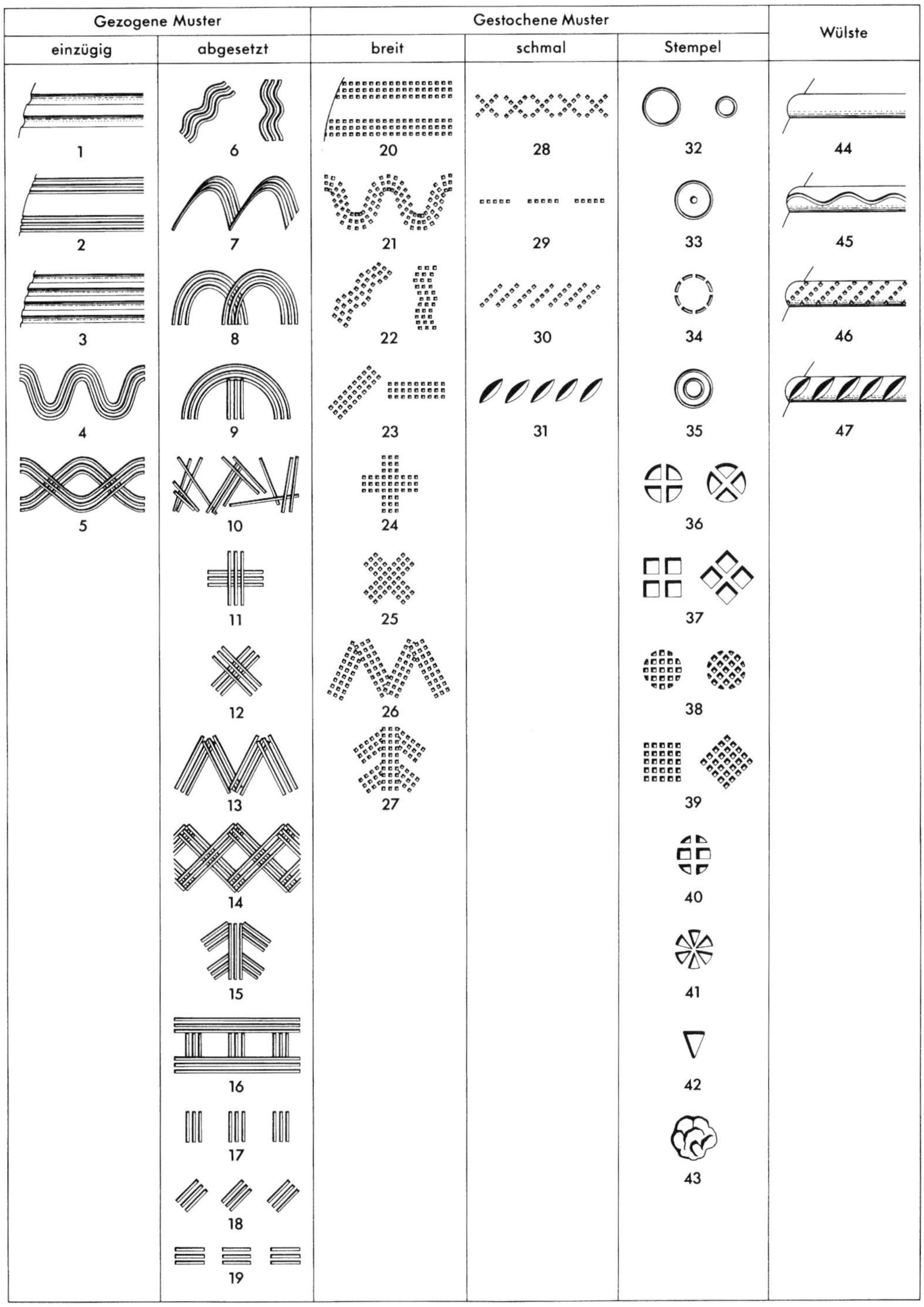

*Fig. 8. Classification system of the decoration on ceramics. The system was originally developed for the material from Scharstorf and has been used on all pottery materials from Groß Strömkendorf (*Meier *1990).*

Grain size distribution analyses

The microscopic investigation also involved a grain size distribution analysis, based on the size and amount of the grains within the silt and sand fractions of the ware. The grain size distribution can be used to identify different sources of pottery production and to indicate different vessel functions. The measurements and calculations were performed by means of image analysing equipment, consisting of a digital camera connected to a frame grabber in a computer. The software KONTRON KS300 was used for the image processing and analysis. It offers interactive measurements and the possibility of automated repetitive measurements according to preset requirements (LINDAHL 2002a, 394).

The results presented in the diagrams include the natural inclusions in the clay as well as the added temper, bearing the risk that a ware consisting of a fine clay with an added temper of crushed rock can be placed into the same group as a ware of a coarse, naturally tempered clay.

This type of analysis is rather new on Slavonic and Scandinavian pottery, so far it has only been performed on ceramics from Scania in Sweden (BRORSSON 2003b) and Wolin in Poland (BRORSSON / STANISLAWSKI 1999).

Thermal colour test (TCT)

A thermal colour test (TCT), developed by B. HULTHÉN (1976), was performed on a sample of pottery sherds. TCT is a method to estimate the firing temperature, based on colour changes of the clay during repeated firing. The Munsell Colour Chart system (MUNSELL 1971) was used as the standard colour reference. The firing was carried out at intervals of 100°C until a temperature of 1000°C is reached. The sample is fired 15 minutes at every interval. Prior to the colour recording, the sherd must cool for at least 15 minutes. The colour is coded according to hue, value and chroma. Each of these data is recorded separately on a graph. It is essential for the success of the test that the colour of the clay only starts changing when the original firing temperature is exceeded. This way, the original firing temperature can be estimated whithin a 100°C interval. A sintering test was also performed (LINDAHL 1986, 128 ff.). This method is primarily used to identify different firing quality groups. Additionally, it can establish the maximum temperature a ware can withstand and still retain an undamaged surface and shape. The type of clay and temper material affects the sintering and melting point. The samples are reheated in 50°C intervals, beginning at 1000°C, up to a temperature of 1500°C. During this process, the condition of the sherd changes from successive darkening to glazing and finally to total melting.

The clays normally found on and near the ground surface in the Scandinavian and North Slavonic areas were formed as sedimentary clays during the last glaciation. Most vessels produced during prehistoric times in this region were made of this type of quaternary clay. Due to several parameters, e.g., a relatively high iron content acting as a fluxing agent, these clays normally start sintering at around 1000°C and melting at around 1200°C. Clays used for the manufacture of Stone ware were usually less rich in iron, formed as a product of *in situ* weathered bedrock (primarily clay). The sintering interval of these clays is much longer when compared to the secondary clays and thereby much easier to control during firing. It is not unusual that this type of clay may be fired to temperatures exceeding 1400°C before it starts to melt.

Shape analyses – Proportional analyses

A proportional analysis is a variant of a shape analysis. It is a method based on studies by G. D. BIRKHOFF (1933) and A. O. SHEPARD (1963) and later developed by H.-Å. NORDSTRÖM (1972). Studies by C. ALEXANDER (1973) also played an important role in the development of the shape analysis.

This analysis is mainly performed on vessels for which the whole profile can be reconstructed. The proportion of the vessel profile is measured and calculated to 100 %. Every change in the shape is plotted by so-called characteristic points onto a diagram. The characteristic points are: the rim point (R), the vertical tangents points (V_n), the inflection points (I_n), the tangent point (T), the corner points (C_n), and the base point (B_n). The distance measurements are converted into percentages of the vessel height, which is set at 100 %. The data, in percentages, are then plotted into a coordinate system in such a way that the vertical symmetry axis coincides with the Y-axis and the base plan with the X-axis.

With this method it is possible to compare vessels of different heights. It can be studied if and how different potters vary in their idea of how to shape a pot for a certain function. Studies on traditional vessel manufacturing have proven that a potter has a very clear idea about the proportion of a vessel, no matter if she/he forms a large or small vessel (LINDAHL / MATENGA 1995, 39 ff.).

The results of the shape analysis can also be connected to other analyses, such as microscopy, TCT, or the study of vessel type and decoration.

History of research of ceramics

The pottery from Groß Strömkendorf mainly comes from the former Slavonic area in northern Germany and Poland. Additionally, there are also ceramics from the Scandinavian countries, as well as from Western Europe, mostly originating from the region near the western part of the river Rhine, the Vorgebirge-Eifel area and the North Sea coast. The history of pottery research in the countries is quite diverse. Shape and decoration have been the major classification attributes in Germany and Poland. Quite to the opposite, the focus on wheel-thrown pottery from Western Europe is on the pottery ware and their provenience.

Pottery research has been influenced by different theoretical and practical approaches. There are only a few thorough studies of Viking Age/Early Medieval pottery in Sweden and Denmark where theoretical archaeology has been considered of greater importance. In Poland the opposite approach has resulted in many detailed studies of pottery from the Early Medieval times. However, pottery research should not be an "either-or" situation. It is important to combine a good theoretical background with a ceramological/archaeological approach when investigating pottery.

In this work, the description of the history of pottery research will be concentrated on that of Slavonic pottery, since 90 % of the pottery from Groß Strömkendorf are Slavonic vessels. Furthermore, the site was located within the Slavonic cultural sphere during the 8th and 9th centuries.

One of the first thorough studies of Slavonic ceramics was undertaken by K. A. Wilde (1939) on the pottery from the trading place of Wolin, in western Poland. In the old part of the town, the thick cultural layers contained enormous quantities of vessel sherds. Wilde tried to connect the ceramics and other artefacts to the stratigraphy in the city. He mainly worked with the rim sherds and compared the pottery with material from other sites in Northern Europe, e.g., Birka, Sigtuna and Lund in Sweden, Vejleby in Denmark and Hedeby in Germany. It should be noted that Wilde mostly used non-Slavonic material for comparison.

In 1955, D. Selling presented her dissertation "Wikingerzeitliche und Frühmittelalterliche Keramik in Schweden", a study of the Viking Age/Early Medieval pottery mainly from the Swedish trading site of Birka, in the Lake Mälaren region (Selling 1955). Selling's work set the foundation for the classification of Viking Age and Early Medieval pottery in Scandinavia. Minor adjustments were presented in the 1980s by other researchers (e.g., Broberg / Hasselmo 1981; Carlsson 1982), but Selling's classification system still is valid today. The study of the Birka pottery was concentrated on vessels from the more than 1100 graves found in the 19th and 20th centuries. Selling recognised some important differences in the material and, according to the ware, divided the pottery into four groups named AI, AII, AIII and AIV *(Fig. 9)*. These groups were later interpreted as originating from different areas; AI from Western Europe, AII from the Slavonic area, AIII from the Finno-Ugric area, while AIV was assigned a local Scandinavian origin. Based on the shape of rims and shoulders, each of these main groups was further divided into subgroups.

Selling not only studied the Birka pottery in her dissertation but also used comparative material from the Mälaren area, from other parts of Sweden, mainly the Kalmar area and Scania, and from other countries, e.g., Germany and Denmark. She was a pioneer as to that she combined typological studies with natural sciences. Selling used, for instance, petrographic analyses of thin sections of pottery and thermal analyses in her characterisation of pottery groups.

Selling's work was surely a milestone for Swedish archaeology and the approach for working with large amounts of ceramic material, but nevertheless it also is an object for further discussion and review. One example is the connection between Scandinavia and Russia and the Baltic countries in the 8th and 9th centuries. Selling recognised some vessels as probably of East Baltic origin, but despite this assumption she chose to classify the vessels into group AIV "Einheimische Keramik" (Selling 1955, 164). The collaboration between Swedish and Russian archaeologists has increased since the end of the Cold War in 1989, which has influenced the research exchange in a positive way. Today we know that the contacts between Sweden and Russia during the discussed period were frequent. Pottery is one artefact category which has been used to understand and discuss these contacts. In several graves at Birka, Ladoga vessels, dated to the 9th and 10th centuries and originating from Russia, were identified

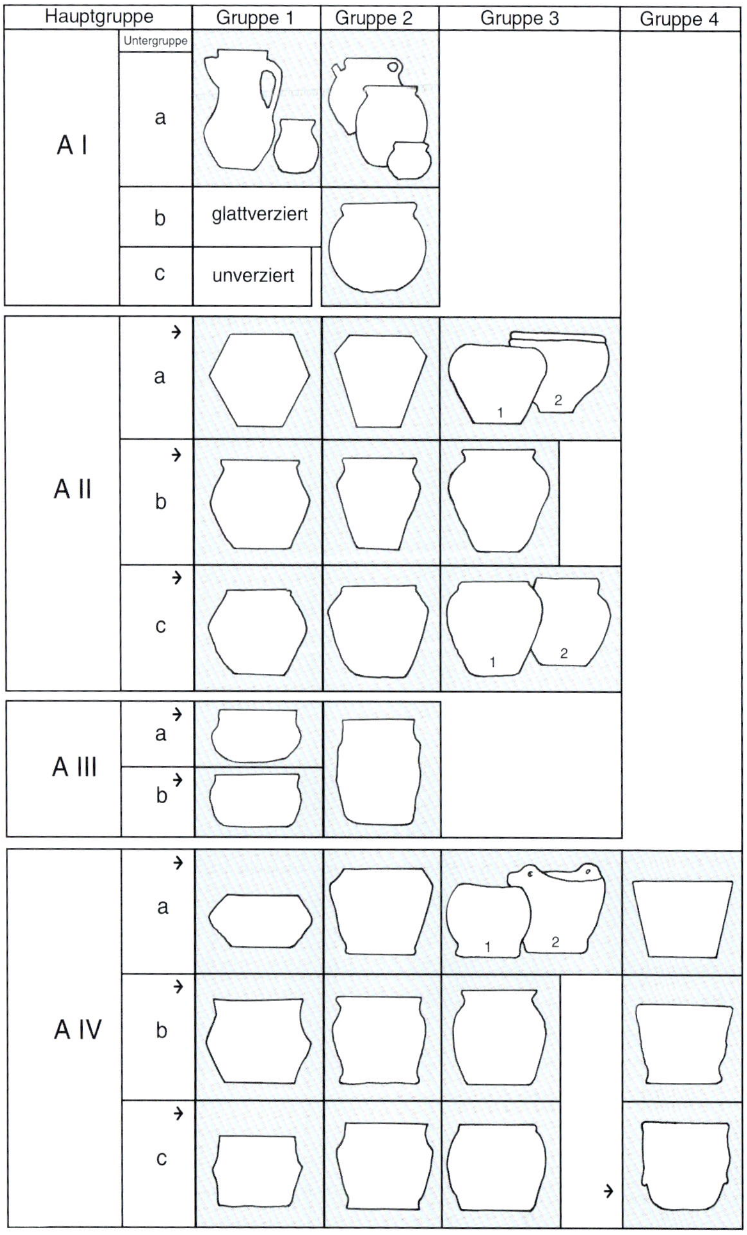

*Fig 9. A classification system on Viking Age/Early Medieval ceramics from different parts of Northern Europe. The system was developed in the 1950s on pottery especially from Birka (*Selling *1955).*

(Brorsson 2003c) *(Fig. 10,1)*. These vessels were classified by Selling as AIV local Scandinavian pottery. The same pattern can be seen in the Birka material from the 10th century. Roslund (2001, 50) has, for example, noticed a vessel of true Russian origin in grave No. 914. This vessel was interpreted as West Slavonic by Selling, originating from the area close to Wolin and the region between the Elbe and Oder rivers. These examples show that no classification system is complete and without discrepancies.

Another important work regarding the classification of Scandinavian pottery from the 8th century are the studies by S. Reisborg (1981a; 1981b) on the pottery from Building group 3 and 4 at the Migration Period site on Helgö, an island in Lake Mälaren. Reisborg did a detailed study of the pottery from the two building complexes, using the vessel shape, decoration and ware. The manner in which the pottery material was processed shows a strong influence from Hulthén's (1974) "On Documentation of Pottery". By looking at different aspects of the sherds, Reisborg tried to establish a chronology for the material and the construction features. The material from each separate building group was divided stratigraphically into an upper and lower layer respectively. The dominating vessel type from Helgö was of a fine or medium-coarse ware, with a flat base and everted rim.

The dominating work regarding the classification of Slavonic pottery is "Die Slawische Keramik in Mecklenburg" by E. Schuldt (1956). It is predominantly a typological study based on shape and decoration of vessels from the 9th to 12th centuries. Schuldt observed some important stylistic differences and classified the material into ten groups, named after the archaeological sites where each pottery group was the most frequent. In this study, Schuldt did not identify Sukow pottery but instead classified it as undecorated Feldberg vessels (ibid. Abb. 20). The different pottery types were put into time spans. For the Feldberg group, the period ranged from the 8th through the 9th century (Schuldt 1956, 22). In addition, the pottery was classified into different "Reihen", which was an attempt to achieve a more precise dating. It is obvious that Schuldt's main aim was not to discuss the role of pottery in society or any other cultural or social aspects, but rather to find a chronology system for Slavonic pottery. In many ways, Schuldt did succeed. However, several vessels did not fit into his system; some could easily be classified as belonging to two different vessel groups.

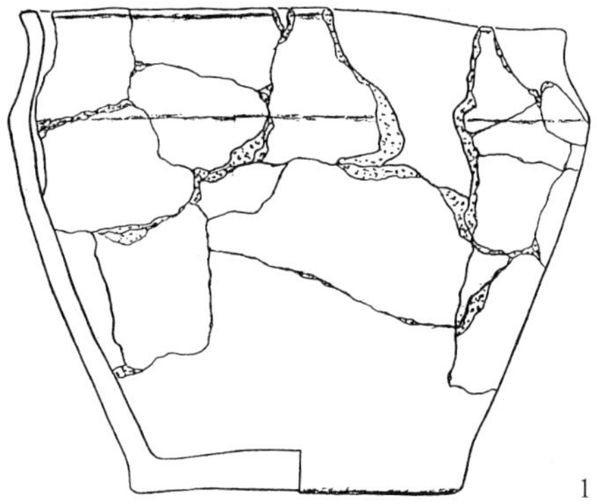

1

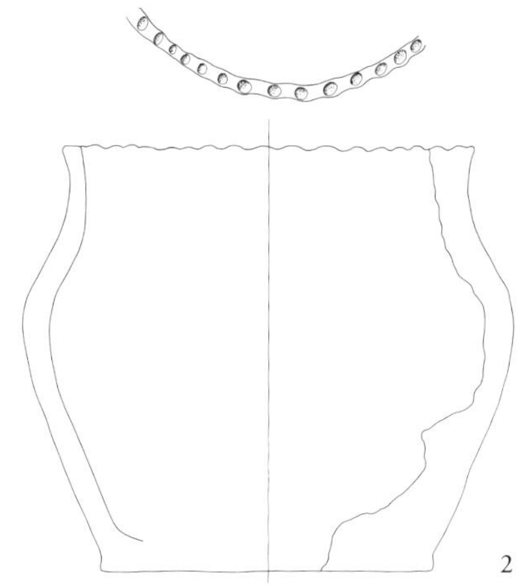

2

Fig. 10. *1 - Pottery of Ladoga type discovered in grave No. 381 at Birka (drawing: A. Plokhov); 2 - Pottery of Curonian type found at Grobina in Latvia (drawing: A. Jeppsson). Scale 1:3.*

Schuldt continued his studies of Slavonic pottery up to the 1980s. He examined the material from one particular site at a time and then put it in a wider perspective. This resulted in several articles dealing with the relation between different vessel categories (Schuldt 1953; 1955; 1963). Besides the Early Slavonic pottery he also worked with younger material, e.g., the pottery from Teterow (Unverzagt / Schuldt 1963).

In his doctoral thesis, T. Kempke presented a partly new approach on how to investigate West Slavonic pottery (Kempke 1981a; 1981b; 1984). He focused on the pottery from the Slavonic hill fort of Starigard/Oldenburg in present-day Schleswig-Holstein. From around 800 to the 12th century, Starigard/Oldenburg was a Slavonic cultural centre of great importance. It was of such importance that in the middle of the 10th century the Archbishop in Bremen decided to send a bishop to the heathen place of Starigard/Oldenburg (id. 1981a, 21). It was not until 200 years later that Starigard/Oldenburg could be regarded as Christian.

Kempke used the stratigraphy at Starigard/Oldenburg to divide the time span from the 8th century to the 12th century into five horizons *(Fig. 11)*. The horizons were then dated by dendrochronology and by artefacts, such as coins, fibulas and combs. Kempke's investigation of the pottery from the hill fort was mainly a typological and chronological work. The most important material were the 2483 rim sherds, which he used as a basis for a classification system into 57 different rim types (Kempke 1981a, 38 Abb. 6). This was in many ways the same method that was used by H. Steuer (1979, 16 Abb. 8,2) on the pottery from Elisenhof. Kempke also investigated the vessel decorations, including 1448 base sherds, of which he examined the shape of the transition between the base and the body. This became an extensive study since Slavonic pottery is normally abundantly decorated. To some extent, Kempke also analysed the pottery ware and divided, for instance, the ceramics' temper into three groups: fine, medium and coarse. In addition, I. Gabriel was involved in the study of the pottery from Starigard/Oldenburg and published a description of the pottery in relation to the different excavations layers (Gabriel 1984).

Mainly based on the decoration and partly on the ware, the pottery was finally classified into different groups named Gruppe U for undecorated pottery, Gruppe P ("Polierte Ware") for burnished pottery, etc. This classification system from Starigard/Oldenburg was then related to the work by Schuldt on the pottery from Mecklenburg. Sukow pottery formed the main part of the oldest phase, which was mainly dated to the 8th century, followed by about 10 % of Feldberg. The Menkendorf and the so-called "Polierte Ware" had a share of about 1 % each of the total material from that phase. The next phase, mainly dated to the 9th century, included the same types but in different frequencies. Especially the amount of Sukow pottery had decreased and that of Menkendorf had increased. The vessel type Groß Raden appeared for the first time during this phase.

F. Wietrzichowski (1989) showed the significance of investigating pottery material from an entire region rather than from just one excavation site. Some years earlier, Kempke (1981a) had shown the importance of comparing local material to other archives, e.g., dendrochronology on wooden remains. In the early 1950s, Schuldt studied pottery from all of Mecklenburg, and in 1990s Wietrzichowski did the same, concentrating on just one special vessel type, the Sukow pottery.

An important work on Polish Slavonic ceramics was accomplished by W. Łosiński and R. Rogosz (1984). Pottery from the Early Medieval city of Szczecin in west-

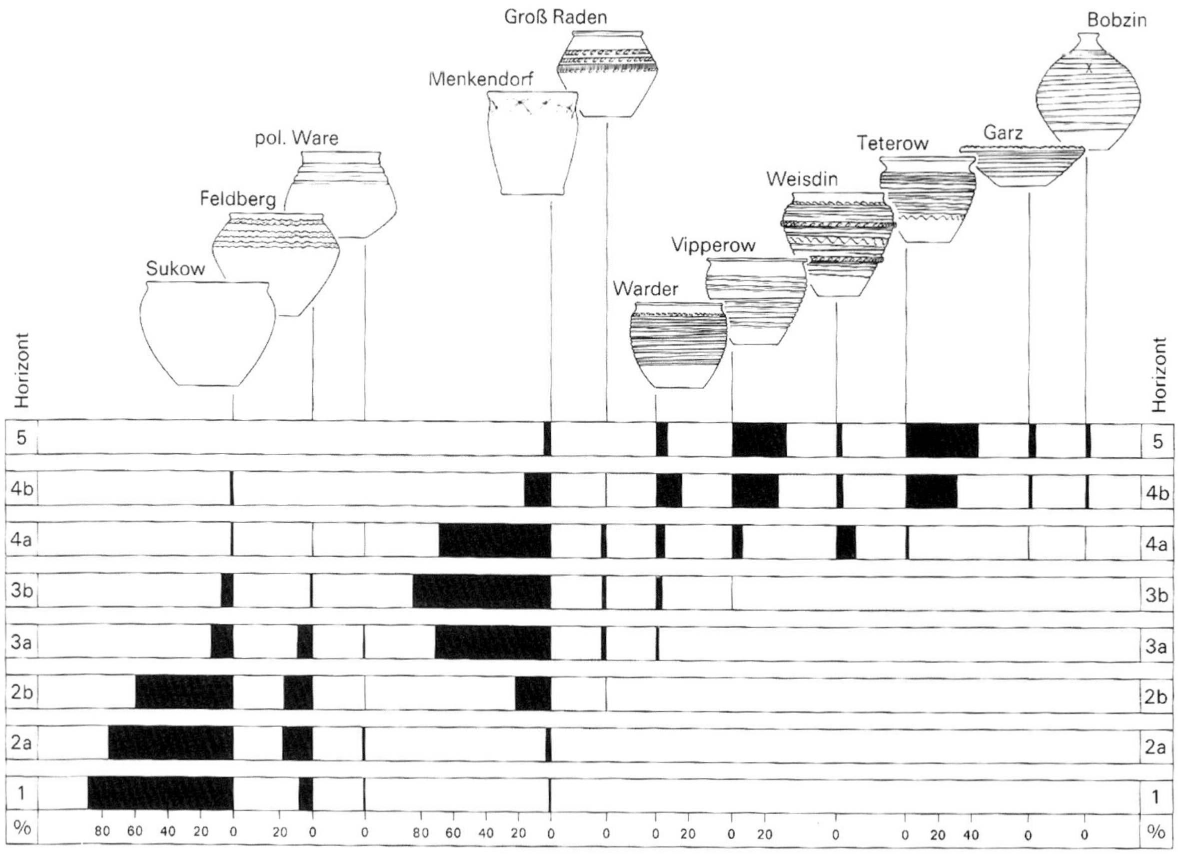

Fig. 11. Datings of the Slavonic pottery appearing in different stratigraphical layers at Oldenburg. Phase 1 is dated from the last quarter of the 7th century to around 800. The other phases cover the time span from 800 to the 12th century (Kempke 1984, Tab. 1).

ern Poland was examined. Instead of using Schuldt's nearly 30 year old classification method, they established a new system, especially designed for the Szczecin material. The pottery was arranged in groups named from A to S. The A group was identical to the Early Slavonic pottery, classified by Schuldt as Sukow. The B and C groups were similar to Feldberg etc. The Late Slavonic pottery was included in groups G to S. In conclusion, the most important difference between the two systems developed by Łosiński / Rogosz and by Schuldt was that the two Polish archaeologists used more detailed groups in their classification. It is, however, quite easy to compare the two classification systems, of which the Schuldt system is still the one used mostly.

S. Brather wrote his doctoral thesis about the Feldberg pottery in northern Germany and Poland (Brather 1996). It was mainly a typological and chronological study, in which Brather, for example, used Kempke's classification system for rim sherds. Although Brather's study was focused on the Feldberg pottery, it also included other Early and Middle Slavonic pottery types, such as Sukow, Menkendorf and Fresendorf.

The work was pioneering for the history of Slavonic pottery in the respect that he used the Feldberg pottery to discuss other aspects than the pottery itself. He did not just examine the development and changes in the making of pottery, but also investigated the relation between the Slavonic groups and the Carolingian kingdom, also involving Slavonic finds from Scandinavia (Brather 1996, 197 ff.).

In his doctoral thesis, M. Roslund (2001) mainly studied the Baltic ware from the countries surrounding the Baltic Sea. He analysed and discussed the dominating components in the contacts across the Baltic Sea. Besides a solid study of the ceramic material, it also contains a thorough theoretical discussion about the history of the Slavonic people and their culture.

Roslund makes an important distinction between the Baltic and Slavonic ware. The latter is defined as vessels made according to Slavonic tradition by Slav people. The Baltic ware, on the other hand, was also based on the Slavonic pottery tradition but produced outside the Slavonic territory from around AD 1000. It also contained some Scandinavian elements (Roslund 2001, 54).

Three other relevant on pottery studies in the Baltic area during the Early Medieval time are "Die Südsiedlung von Haithabu" by H. STEUER (1974), "Die Einheimische Keramik von Haithabu" by K.-D. HAHN (1977) and "Die Importkeramik von Haithabu" by W. JANSSEN (1987). These studies are particularly interesting for the comparison with the Groß Strömkendorf material since, according to the written sources, the traders moved from *Reric* to Sliasthorp (Hedeby).

STEUER (1974) concentrated on the pottery material from the trading area and the cemetery. The pottery consisted mostly of locally hand-made vessels. He also examined various imported ceramics, such as Badorf, Muschelgrus and Pingsdorf pottery. Using relative and absolute datings, he tried to date the ceramic material and the construction features. He mapped out the Early Medieval pottery in the region by comparing the Hedeby material with pottery from other sites on the North Sea coast and in Schleswig-Holstein.

HAHN (1977, Abb. 19) scrutinised the hand-made and local ceramics from Hedeby. The pottery during the oldest phase, dated to the 8th century, was dominated by "Kugeltopf" and "Eitopf". The Baltic ware only amounted to a few percentages of the total material. The investigation of the hand-made ceramics showed that the pottery from the hinterland dominated the household material at the trading site.

JANSSEN's (1987) study of the Hedeby pottery was focused on the imported vessels, to such an extent that would have made it nearly impossible to involve the locally hand-made pottery as well. Janssen's pottery classification was mainly based on the ware and decoration. Most of the vessels were imports from Western Europe, for example from the Rhine area and from Andenne in France.

H. J. MADSEN (1991) compiled a kind of state-of–the-art system for ceramics information in Denmark. Madsen scrutinised several articles dealing with Viking Age pottery. He discussed its role in Denmark and its use for social interpretation, showing the spread of different vessel types. Madsen noticed that Jutland in western Denmark differs from Zeeland in the eastern part of the country due to different contact networks.

Another relevant study for the Groß Strömkendorf material is on ceramics from the Danish city of Ribe, undertaken by C. FEVEILE and S. JENSEN (2000). Due to an informative stratigraphy and objects dated by dendrochronology, it was partly possible to divide the 8th century layers into periods of 20 years. The ceramics are of similar types as in Groß Strömkendorf.

An outstanding reference work on medieval pottery in Northern Europe is the "Handbuch zur mittelalterlichen Keramik in Nordeuropa" (LÜDTKE / SCHIETZEL 2001), consisting of three volumes with about 550 pages each. Several authors, specialised in different ceramic topics, have contributed to themes such as dating, technology and geographical distribution. The work includes a basic terminology for medieval pottery in northern Germany (ERDMANN ET AL. 2001), which is important since a variety of terminologies is used in different countries or even by different institutions in the same country, constituting a major problem in the archaeological sciences.

The history of research concerning West European pottery is quite different from the North European tradition. Although pottery from the Rhineland and the Eifel area has frequently been found on sites in northern Germany, Denmark, Sweden and Norway, it did not have the same influence as in Northern European archaeology.

The works used mostly in Northern Europe are Janssen's study of the Hedeby pottery and that by W. A. VAN Es and W. J. H. VERWERS (1980) of the Dorestad pottery, for which a new classification system was established based on criteria such as manufacturing technique, colour of the vessel surface, coarseness of the ware, and vessel shape and decoration. The pottery found at Dorestad derived mostly from the Rhineland and the Eifel area.

As pointed out by C. KELLER (2004a), the Rhineland pottery has been discussed and analysed in terms of long-distance trade to centres such as Dorestad and Hedeby. The different pottery categories have been objects of interest outside their production area, but not in the region where they were actually produced. Another example are SELLING's (1955) different works, mainly on the Badorf and Tatinger pottery. Her interest in the vessel types was surely affected by the finds in the graves at Birka.

Despite the fact that it has been known for more than 100 years that Carolingian pottery was produced in the area, the research interest on the pottery from the Rhine and Eifel area has been low. One example from the 1960s is the find of three pottery kilns in Brühl-Eckdorf, near Cologne (JANSSEN 1970). Since the end of the 1990s, more and more material from the Rhineland and Eifel has been published (e.g., KELLER 1998; SANKE 1999), allowing further investigations of West European pottery found on sites in the Baltic area. A proper local classification system based on dendrochronology would also be useful.

Another type of ceramics that appears during the 8th century is most likely from present-day France, mainly dated to the Merovingian period. As mentioned by M. REDKNAP (1998, 61), the research about this pottery ware is even more limited than that on the ceramics from the Rhineland and Eifel.

Investigations on the shell-tempered Muschelgrus pottery, originating from the Frisian area, have had a neglected role in archaeological studies. It is chiefly the material from Dorestad (VAN Es / VERWERS 1980), Emden (STILKE 1995a), Elisenhof (STEUER 1979) and Hedeby (JANSSEN 1987) that was examined in detail, with the focus on the specific temper material. With a few exceptions (e.g., Elisenhof and Emden), the shapes did not attract much interest. H. STILKE's (1995b, 15 Fig. 4) article about shell-tempered pottery from the Frisian coastal area includes discussions concerning the ware, shapes and distribution of shell-tempered vessels in Northern Europe.

The English Shelly ware covers several decades of the pottery tradition on the British Isles. Its groups are more complex than those for the shell-tempered pottery from the Continent, which is shown by a wide range of wares and forms. The production of Shelly ware seems to have been restricted to eastern and southern England (McCARTHY / BROOKS 1988).

A significant part of the study of the pottery from Groß Strömkendorf is the ceramological analysis of the material. As mentioned above, Selling was one of the first archaeologists in Scandinavia to integrate analyses of clay and temper – the ware – into her investigation. But more than 20 years earlier, the Norwegian archaeologist J. BØE (1931, 204 ff.) carried out chemical analyses on Iron Age pottery from Bergen. Before the 1970s, analyses of ware, clays and minerals in Scandinavia were made on request by a geologist or chemist. The dialogue between the geologist/chemist and the archaeologist was often very poor. It was difficult for the archaeologist to extract the useful results from the figures and tables.

HULTHÉN (1977) presented her thesis on Neolithic and Bronze Age pottery from southern Scandinavia. Some years earlier, Hulthén had published several books and articles concerning different methods on how to approach large pottery assemblages (cf. ID. 1974; 1976). The new approach to the ceramic studies in Scandinavian archaeology was that Hulthén was an archaeologist who had also studied geology and statistics, and this combined competence was very fruitful.

A. LINDAHL (1986) further illustrated the importance of laboratory analyses when evaluating ceramic material in his thesis on medieval pottery from Dalby, outside Lund in Sweden. This work is a methodological study on how to apply different types of analyses to "anonymous" sherd material. In O. STILBORG's (1997) thesis on the pottery from the Gudme-Lundeborg area on Funen in Denmark, pottery craft from the Late Roman Iron Age was compared to contemporaneous sites within a limited area, mainly the settlements of Gudme and Brudager, the cemetery at Møllegårdsmarken and the trading site at Lundeborg.

One of the most common scientific methods for analysing pottery is petrographic microscopy of ceramic thin sections. Other methods, such as Thermal Colour Test, thermoluminescence, lipid analyses and magnetic susceptibility have also been applied in Scandinavia. Since the 1990s, the analyses have been used more and more to discuss the craft and the process of development rather than questions such as if the pottery was produced locally or if it was imported.

The natural science research on pottery in Poland, Germany and the Netherlands has significantly increased since the 1970s. The prevailing method is the petrographic study of thin sections. Neutron activation analyses as well as chemical analyses have frequently been used on material from Germany and Poland. Several different analyses of pottery sherds from the Rhineland-Eifel area have been compared with pottery from other sites, e.g., Dorestad (BARDET 1995). With very few exceptions, the analysis has focused on where the vessels were produced (e.g., BUKO / PELA 1997). Although this is a fundamental question, it is important to remember that there are other aspects to the pottery and the craft as well.

A study involving thermoluminescence (TL) and magnetic susceptibility (MS) of wheel-thrown pottery from the Danish city of Ribe showed that coarse wheel-thrown pottery was made in the city during the 8th century (FEVEILE ET AL. 1998). The same type of analysis has been performed on other material, but these results have not been clearly defined. One problem, which became apparent when the samples previously analysed by the TL/MS method were examined under the microscope (thin sections), was that the analyses of magnetic susceptibility and thermoluminescence were not only performed on the clay, but also on the temper material (BRORSSON 2005, 313). Thus, rock temper rich in ore will affect the magnetic susceptibility and may perhaps indicate a foreign origin of the vessel, when in fact it represents another craft tradition, vessel function or a different random choice of temper material.

Early Medieval pottery types in the Baltic region

Slavonic pottery

Research on Slavonic pottery has been mainly based on the study of shape and decoration. It goes back to the middle of the 19th century, when different studies laid the foundation to later investigations (KEMPKE 1984, 12 ff.). Based on the Wolin material, K. A. WILDE (1939) made one of the first detailed studies of the West Slavonic pottery. This work was followed by other investigations. Before the 1980s, E. SCHULDT's (1956) research on the pottery from Mecklenburg was the most important and also most widely used. Others are, as mentioned above, T. KEMPKE's (1981a; 1981b) study of the Oldenburg material and S. BRATHER's (1996) research on the Feldberg pottery. A contribution to the dating of the Slavonic pottery is presented in the work "Handbuch zur Mittelalterlichen Keramik in Nordeuropa". A special chapter in book 2 is dedicated to coin-dated ceramics (STILKE 2001, 763 ff.). The earliest coin dating of Slavonic pots is from the 10th century.

Unlike Scandinavian pottery, the Slavonic vessels have miscellaneous and complicated decorations. The most common decoration types are vertical lines and wavy-lines in different compositions. Additional decorations are impressions and plastic ornaments. However, the earliest West Slavonic type, the Sukow, is not decorated, which may indicate an influence by its precursor, the undecorated Prague pottery. The first decorated West Slavonic pottery is Feldberg. It has also been suggested that undecorated Feldberg pottery was made already before the second half of the 8th century (WIETRZICHOWSKI 1989, 77). The interpretation was based on a find of Feldberg pottery together with a Scandinavian bird fibula at Schönfeld, Kr. Demmin. The fibula was dated to 650–725.

The Slavonic vessel shapes differ from Scandinavian types. The dominating shape are bi-conical pots with everted rims. Only minor differences in shape between vessels from the 8th and 10th centuries can be observed. The differences depend mainly on technical developments within the ceramic craft. One example is the design of the rim. During the 8th century, the dominating rim shape was slightly rounded, whereas during the late 10th century the most common rim shape was more or less cut off, forming a square profile. Several archaeologists have noticed this major difference; T. KEMPKE (1981a; 1981b) developed a classification system, based, among other things, on the shape of the rim.

The Slavonic pottery group consists of several different vessel types, dating to the late 7th to the 13th century. West Slavonic pottery can be divided into 13 groups, according to Schuldt's classification and new types added by Kempke. However, borders between, for instance, the Sukow, Feldberg, Menkendorf and Fresendorf pottery are quite diffuse and under discussion (KEMPKE 2002, 237). Since the work on the Groß Strömkendorf material is focused on the 8th and 9th century, a short presentation of the pottery groups from this period follows below.

Sukow (Plates 1–8)

Sukow pottery belongs to the Early Slavonic pottery. It is recognised by the S-shaped profile and the undecorated vessel surface *(Fig. 4,1)*. The rim is everted, the shoulder is normally fairly distinct and the base is flat. The vessels are coil-built by hand. When Schuldt wrote his work on the pottery in Mecklenburg, he did not include the Sukow pottery. He only discussed the undecorated Slavonic pottery later in his study of the material from Groß Strömkendorf (SCHULDT 1955, 137). According to the technical quality of the ware, Schuldt interpreted it as an Early Slavonic type and assumed that it was some kind of precursor to the Feldberg and Fresendorf pottery.

Another paper by SCHULDT (1963) concerned the relations between the undecorated Slavonic pottery and the Feldberg group. He noticed important similarities in the vessel profiles and the shape of the rim. The parallels to the undecorated Prague pottery from the Czech Republic were clear. He dated the vessels in Mecklenburg to the 7th to 9th century (ID. 1964, 258 ff.).

WIETRZICHOWSKI's (1989) *Diplomarbeit* was on Sukow pottery in Mecklenburg. As mentioned above, Wietrzichowski studied the ceramic material from a total of 387 places in the region. He classified the Sukow pottery and the undecorated Feldberg pottery into four shape groups, divided into 45 different types of rims *(Fig. 12)*. It was mainly the length of the rim and the shape of the lip that were used for the classification of the vessel material. Shape group 1 and 2 were reserved for coarse

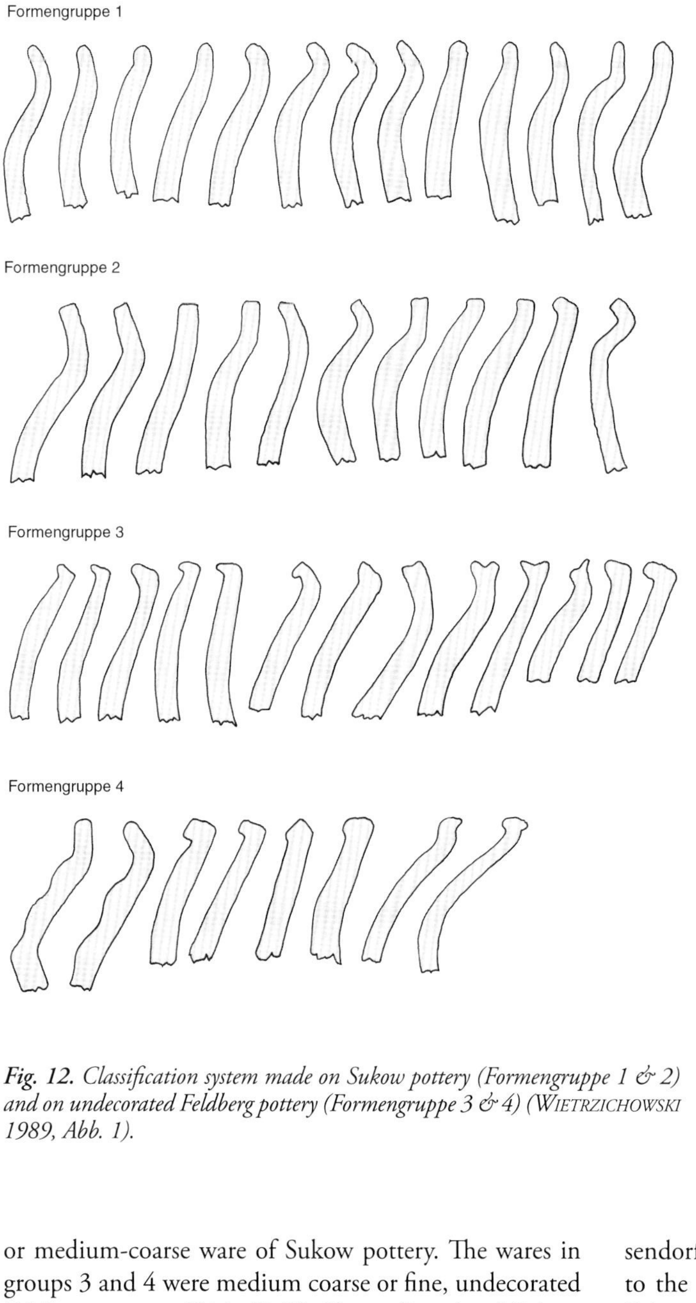

Fig. 12. Classification system made on Sukow pottery (Formengruppe 1 & 2) and on undecorated Feldberg pottery (Formengruppe 3 & 4) (WIETRZICHOWSKI 1989, Abb. 1).

or medium-coarse ware of Sukow pottery. The wares in groups 3 and 4 were medium coarse or fine, undecorated Feldberg pottery (ibid. 40 ff.). The earliest dated Sukow pottery in Mecklenburg is from the middle of the 7th century. The transition into Feldberg pottery may have started after this period and continued, according to Wietrzichowski, through the period of the undecorated vessels in shape groups 3 and 4. It is recognised by the fact that the younger pottery consists of stroked rims, finer ware and is well fired.

Sukow pottery has been found in the whole of Mecklenburg-Vorpommern (BRATHER 1993, 327 Abb. 3), but predominantly in the southern and western part of the region *(Fig. 13)*.

The Sukow pottery is very rare outside the West Slavonic region. The lack of decoration makes it difficult to differentiate it from other pottery, e.g., the Saxon and Scandinavian material. According to the literature, only undecorated complete Slavonic vessels were found outside the West Slavonic area. In the cemetery at Birka, Sweden, two vessels of Slavonic pottery with short rims, classified by SELLING (1955, 126) as AII:3c1, were discovered. There are also three undecorated vessels classified as AIV:3b, which may be interpreted as Slavonic (ibid. Taf. 54,7–9). However, these five vessels from Birka may instead be Sukow. Sherds from Sukow vessels were also found in the cultural layer “the Black Earth” at the trading site (BÄCK 1995, 10). The dating of the material in Birka covers a wide time span, stretching from the 9th to the 12th century (SELLING 1955, Abb. 63). At the Viking Age trading site of Åhus in eastern Scania, sherds of Sukow pottery were found (Lund University, History Museum 30192), most likely belonging to the 8th century.

At the hill fort at Oldenburg, Sukow pottery appeared at the beginning of the activity of the settlement, e.g., in Horizon 1 in the 8th century (KEMPKE 1984, 35 ff.). In all likelihood, the production stopped during the 9th century, though sherds of Sukow pottery were discovered in much younger layers.

Sukow pottery was also present at the trading place of Wolin in Poland. Sherds of this type appeared in the excavation of the harbour area in the stratigraphically lowest layers (BRORSSON / STANISLAWSKI 1999, 290 Tab. 1, 2), which were dated to the second half of the 8th century. The total share of Sukow pottery in the layer is about 2 %. The rest of the material consists of Feldberg, Fresendorf and Menkendorf pottery. In the structures dated to the second half of the 9th century, the occurrence of Sukow pottery almost vanished (ca. 0.4 % of the total material).

Nearly the same pattern regarding dating and the amount of Sukow pottery can be noticed in the material from Szczecin, south of Wolin. The earliest finds are dated to the second quarter of the 8th century. The pottery type decreased rapidly and almost vanished during the second half of the 9th century (ŁOSIŃSKI / ROGOSZ 1984, 202 Abb. 17).

Feldberg (Plates 9–14)

Feldberg belongs to the Early Slavonic pottery group. It was classified by SCHULDT (1956, 19 ff.) in his work on

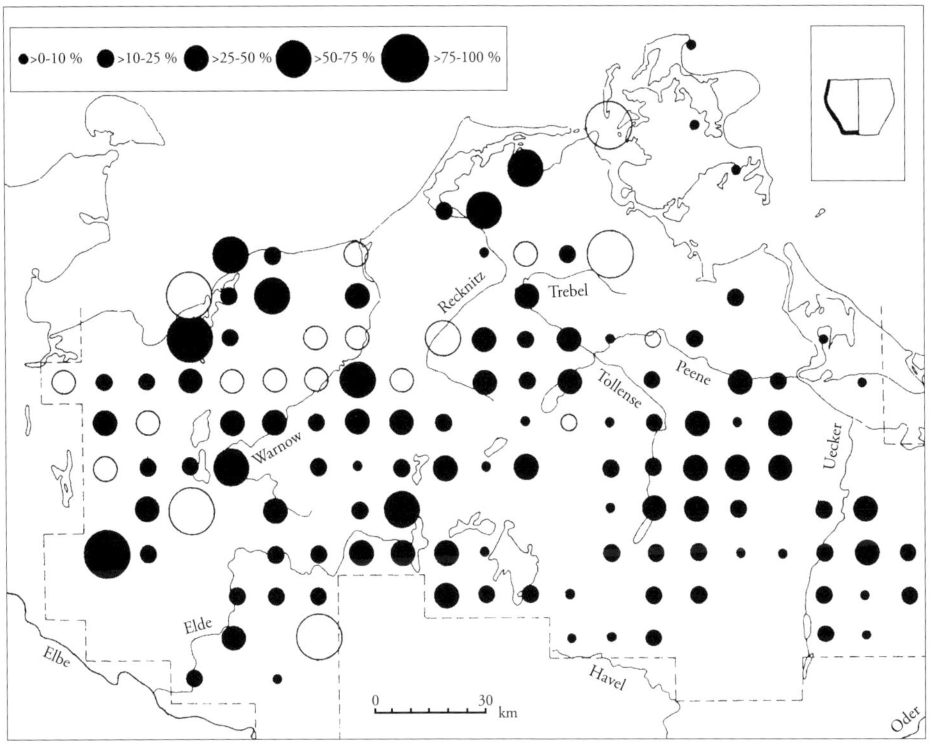

Fig. 13. Distribution of Sukow pottery in Mecklenburg. A hollow circle means uncertain amount (BRATHER 1993, Abb. 3).

the pottery from Mecklenburg. According to Schuldt, most of the vessels in the group are decorated but he also discussed the problems with undecorated Feldberg pottery. It was especially the pottery from the site of Sukow that caused problems to his classification system (ID. 1963). He finally classified Feldberg pottery as an independent type of decorated Slavonic vessels. The vessels were built up by hand and shaped on the turntable. The shape is similar to Sukow pottery, which features an S-shaped profile and everted rims. The lip of the rim is mostly stroked off and the base is flat. Different kinds of plastic decorations appear, e.g., knobs or cordons. The rim is often decorated with different comb impressions or short lines. Generally, Feldberg has been considered to be of a slightly finer quality than Sukow pottery.

WIETRZICHOWSKI's (1989) study on the Sukow pottery and the undecorated Feldberg pottery from Mecklenburg shed new light on the discussion of the relationship between the two vessel groups. The development from Sukow into Feldberg types came quite early, after the second half of the 7th century (ibid. 77). Wietrzichowski used the name "Formengruppe 3 & 4" for the undecorated Feldberg pottery *(Fig. 12)*.

BRATHER (1996) studied Feldberg sherds from a total of 1250 sites in Northern Europe. According to Brather, the transition from undecorated Slavonic types into decorated types, e. g., Feldberg, was under the strong influence by the fashion of using Frankish Badorf pottery in the household (ibid. 210). Furthermore, the Feldberg pottery was mostly found in the eastern part of Mecklenburg (ibid. 199) *(Fig. 14)*.

In Scandinavia these ceramics have been found at many sites (CALLMER 1988, 673 Abb. 4), of which the most well-known is Birka. Here, a number of Feldberg vessels were found in several graves and in the cultural layer, "the Black Earth" (SELLING 1955; BÄCK in prep.). Another important site with finds is Åhus in Scania (CALLMER 1991, 41).

Without exception, other places in Scania with finds of Feldberg pottery are located within a few kilometres from the sea, which is a good indication of trade by boat (BRORSSON 2003b; 2003d; 2006). Microscopy analyses have shown that nearly all Slavonic pottery was imported. Only some vessels were most likely produced locally near the trading or market sites in Scania.

In Mecklenburg, this pottery appeared in the 8th century and disappeared during the 10th century (SCHULDT 1956, 21 f.). A more precise dating for its appearance in Mecklenburg is still missing. At Starigard/Oldenburg in Schleswig-Holstein, Feldberg pottery was found in Horizon 1 and 2, dated from the 8th to the 9th century (KEMPKE 1984, 35 ff.).

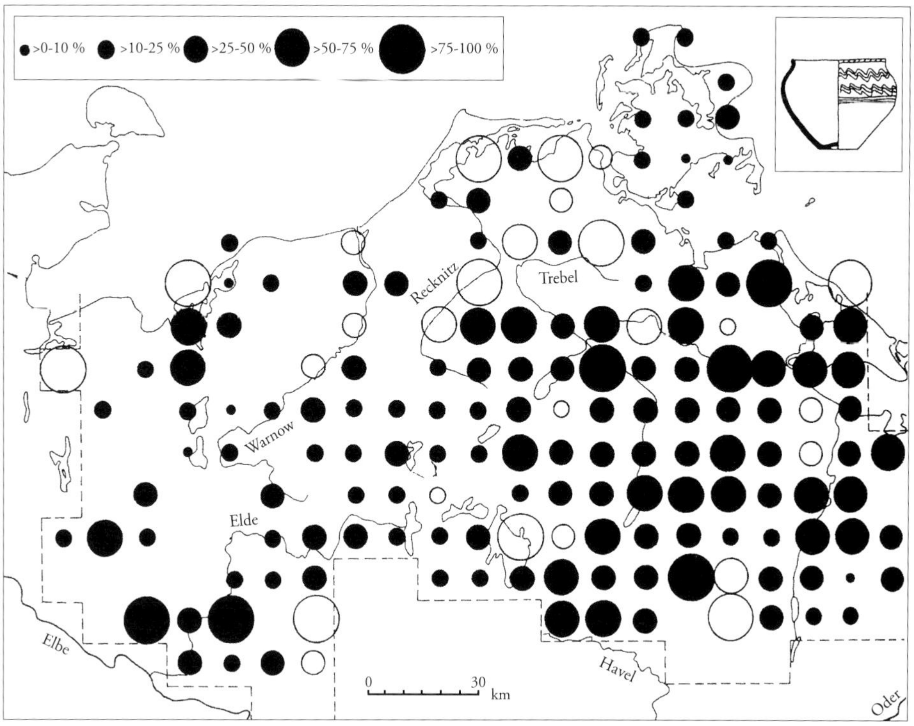

Fig. 14. Distribution of Feldberg pottery in Mecklenburg. A hollow circle means uncertain amount (Brather 1993, Abb. 2).

One of the sites where Feldberg pottery first appeared is Wolin. A large amount of these vessels – about 80 % of the total material – were recovered already in the oldest layer of the excavation at the harbour area, which dates to the late 8th century (Brorsson / Stanislawski 1999, 290 Tab. 1, 2). The excavation has shown that Feldberg pottery began to decrease during the second half of the 9th century and nearly disappeared in the next phase, the first half of the 10th century.

The earliest finds of this pottery from the city of Szczecin are dated to the second quarter of the 8th century (Łosiński / Rogosz 1984, 202 Abb. 17). It was the dominating vessel type in this period and during the whole 9th century. Its share decreased rather quickly and fell below 10 % of the total material after the end of the 10th century.

Fresendorf (Plate 15,1–8)

Fresendorf belongs to the Middle Slavonic pottery group. It has several similarities with the Feldberg pottery, sometimes with only subtle differences (Schuldt 1956, 25 ff.; 1980, 12). The Fresendorf pots were coil-built by hand and thereafter finally shaped on the turntable. The vessels were usually abundantly decorated with straight and/or wavy lines. The dominating rim shape is inverted. The base is flat and the profile of the vessels is more or less like a wide barrel.

Fresendorf pottery has mainly been found in the eastern part of Mecklenburg, including the island of Rügen (id. 1956, Abb. 43) *(Fig. 15)*. Roslund (2001, 248 f.) commented on these facts in his work on the Baltic ware, dated from the late 10th to the 13th century. He argued that the Fresendorf pottery, together with local Scandinavian pottery, was the precursor to the Baltic ware in Scania. The Baltic ware at sites such as Lund is dominated by vessels with rims bending inwards. It is likely that the Slavonic tribe, the Ranes, which had control of the island of Rügen, influenced the shape of Baltic ware in Scania and Sweden. One of Roslund's hypotheses was that the Slavonic people were taken as slaves, brought to Scandinavia and made to perform different daily chores, of which making pottery was an important one.

In his study of Early and Middle Slavonic pottery from Mecklenburg-Vorpommern, Brather (1993, 330 Abb. 5) confirmed that Fresendorf was the dominating pottery type on the island of Rügen. In Mecklenburg, it is mainly dated to the 9th and 10th century (Schuldt 1953, 145 f.).

Fresendorf pottery was also found at the trading site of Wolin, where a large amount of sherds was identified during the excavation of the harbour area. The earliest finds date to the second half of the 8th century. Their amounts were the highest during the first half of the 10th

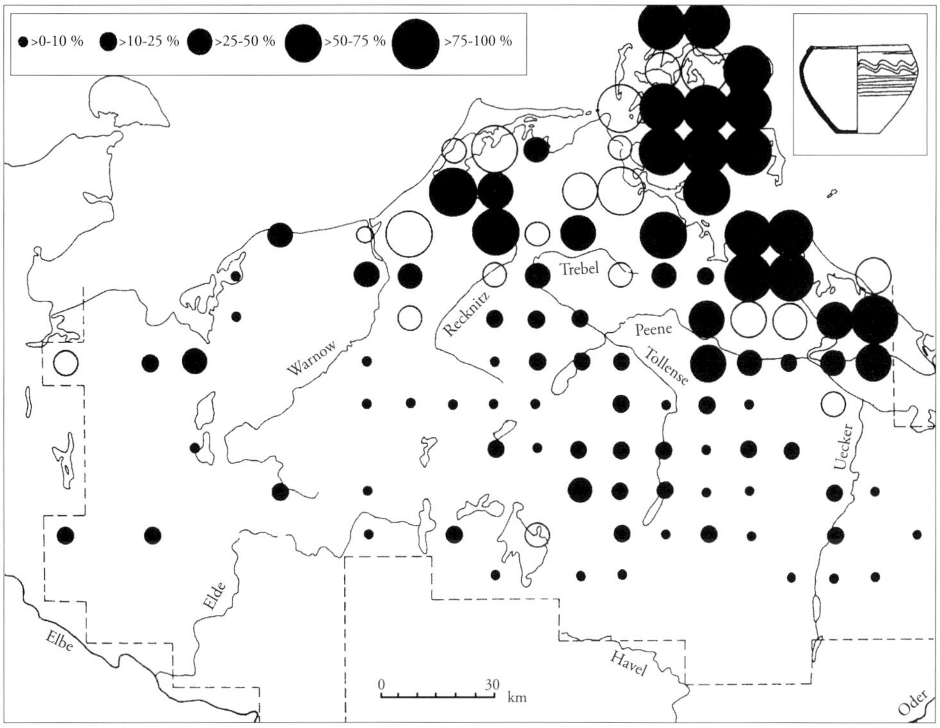

Fig. 15. Distribution of Fresendorf pottery in Mecklenburg. The connection to Rügen and the Ranes is obvious. A hollow circle means uncertain amount (BRATHER 1993, Abb. 5).

century and started to disappear during the same century (BRORSSON / STANISLAWSKI 1999, 291 Tab. 2).

In the study of the material from the city of Szczecin, Fresendorf pottery is classified as type F. It is not common in Szczecin; the first finds derive from the middle of the 9th century (ŁOSIŃSKI / ROGOSZ 1984, 203).

The fact that Fresendorf pottery seemed to be dominating in the north-eastern part of Mecklenburg is interesting, bearing in mind that Kempke did not classify any sherd as belonging to the Fresendorf group in his studies of the Starigard/Oldenburg (KEMPKE 1984, 68) and Hamburg (ID. 2002) pottery. This pottery was, together with a special type of Feldberg, classified as "Wulstrandtöpfe, Gruppe W". The few finds of such ceramics in Starigard/Oldenburg are rather connected to the Feldberg types.

Outside northern Germany and Poland, Fresendorf pottery has been found on several locations where Feldberg pottery also occurred (BRORSSON 2003b; 2003d). In Scandinavia, Feldberg pottery was more common than Fresendorf, according to the available publications. The reason for this is uncertain. However, differences in the chronology of these sites may have affected the quantity of the finds of Feldberg and Fresendorf ceramics. The spread of Fresendorf outside its core area certainly was dependent on the group of Slavonic people that travelled westward and northward respectively.

Menkendorf (Plates 15,9–10 and 16)

Menkendorf belongs to the Middle Slavonic pottery group, also classified by SCHULDT (1956, 9 ff.). It is characterised by biconical, S-shaped profiles, long necks and everted rims. The shoulder is located much higher than on the Sukow and Feldberg pottery. The base is flat. Menkendorf pottery is decorated with different types of straight lines, wavy lines and impressions (mostly comb impressions), especially on the neck. The plait pattern is a common type of decoration *(Plate 16,4)*. The pots were coil-built by hand and shaped on a turntable.

In Mecklenburg, Menkendorf pottery made its first appearance in the 8th century, but its main occurrence was in the 9th century (ibid. 14 f.). The Menkendorf vessel type is still common in features dated to the 10th century. At Groß Raden, it dates from the beginning of the settlement in the middle of the 9th century to the end of the 10th century (ID. 1981, 39).

In Oldenburg, Menkendorf pottery was discovered mainly in layers dated to the 9th and 10th centuries, but it also appeared in other periods (KEMPKE 1984, 90 Tab. 1).

Quantitative studies of Menkendorf pottery from Mecklenburg-Vorpommern show that the type is to be found in the middle and southern part of the region (BRATHER 1993, 329 Abb. 4) *(Fig. 16)*. Some finds were

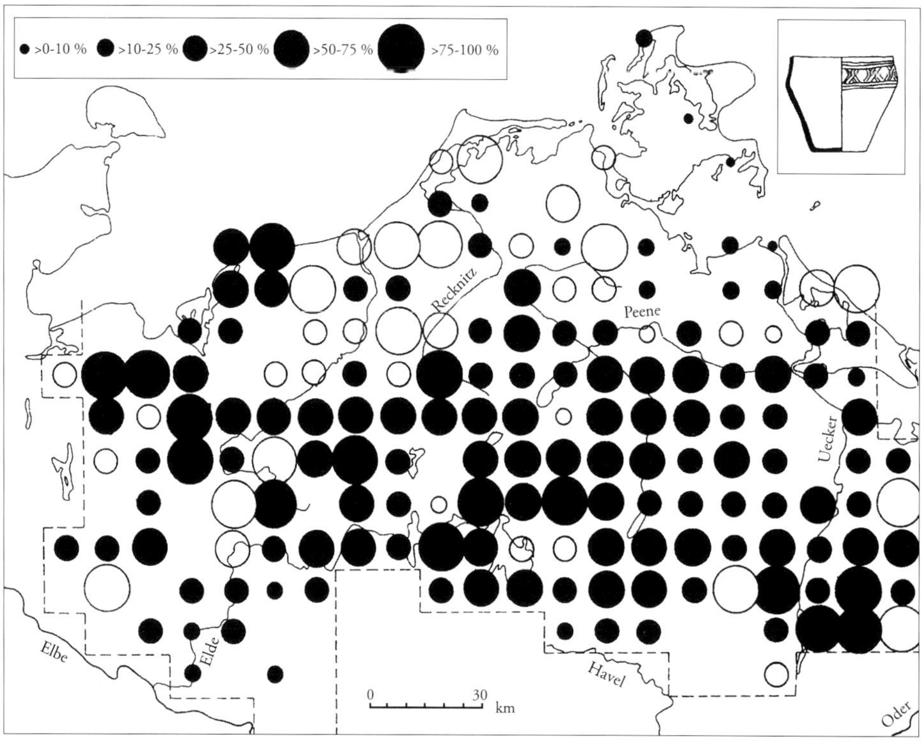

Fig. 16. Distribution of Menkendorf pottery in Mecklenburg. The ceramics have been found in rich amount all over the region, except in the north-eastern part. A hollow circle means uncertain amount (Brather 1993, Abb. 4).

made in the western part of Mecklenburg, close to the coast of the Baltic Sea. The Menkendorf pottery has been connected to the region in Mecklenburg occupied by the Slavonic tribe of the Obodrites (Schuldt 1956, 14 ff.).

In Wolin, this pottery was found in, for example, the excavations of the harbour area. The vessels started to be used during the second half of the 8th century and seem to have disappeared during the 10th century (Brorsson / Stanislawski 1999, 290 Tab. 1, 2). They reached their peak in the second half of the 9th century and the first half of the 10th century.

Menkendorf pottery from the city of Szczecin is classified as type D. Its first appearance is dated to the first quarter of the 9th century but only in quantities of less than 10 % of the total material. However, it dominated the pottery material at the early part of the 10th century (Łosiński / Rogosz 1984, 202 Abb. 17).

Menkendorf is not as common as Feldberg and Fresendorf pottery in Scandinavia, according to the available literature. The only known places with finds of Menkendorf pottery in Sweden are Mölleholmen (Kelm 2000, 29), Löddeköpinge (Brorsson 2000, 199 f.), Ystad (Strömberg 1978, 75) and Järrestad (Brorsson 2003b, 355) in Scania. The pottery from Mölleholmen and Löddeköpinge consists of quite uncertain finds; it could have been mistaken for the later Baltic ware. The trading site at the quarter Tankbåten in Ystad, where a few Menkendorf sherds were identified, was dated to the 8th and 9th centuries. The settlement at Järrestad differs from the other Viking Age sites in Scania. The place has been interpreted as a manor and estate for the Jarl (Earl) controlling the region (Söderberg 2003, 158; 2005). The Slavonic pottery at Järrestad was, in relation to the finds at Wolin and Oldenburg, dated to the 9th century.

At least seven locations with finds of Menkendorf pottery are known in Denmark (Brather 1996, Liste 2). Among them are Jutland, Sealand and Bornholm. Menkendorf pottery was found, for example, in a pit inside the fortress Trelleborg, near Slagelse on Sealand (Nørlund 1948, Fig. 108a) and in a pit house at Riberlund in Ribe on Jutland. The house was dated to the 9th century (Claus Feveile, den antikvariske samling, Ribe, pers. comm.). One find was also made on the island of Bornholm, in a coffin grave at Runegård on the southern part of the island (Watt 1983, 147 Fig. 10). Based on the relation to other finds, the grave was dated to the second half of the 10th century. The same dating was obtained for the fortress Trelleborg.

Scandinavian pottery *(Plate 21)*

Viking Age Scandinavian pottery is a group of pottery that has not been of much interest to Scandinavian archaeologists. The Viking Age has been regarded as a period where the pottery craft more or less degenerated (SELLING 1971). The typical undecorated, comparatively uniform vessels of coarse ware have thus not evoked much attention. The first proper investigation of vessels from that period was undertaken by D. SELLING (1955). Selling worked mainly with material from the trading site of Birka and discussed the imported and local vessels. The AIV group was interpreted as indigenous to Scandinavia. This group was further divided into 13 subgroups, based on the rim and body shape (ibid. Abb. 1). Selling also tried to correlate vessel types with different periods from the 8th to the 12th century, and concluded that some types were actually in use throughout the whole period. The local group AIV is mixed with foreign pottery, which is a major problem when using Selling's classification system. Apparent West Slavonic pottery can be found among the domestic group, and a major eastern influence is also evident at Birka. Ladoga pottery, originating from the region near Staraja Ladoga and Ryurikovo Gorodishche in western Russia, was identified (BRORSSON 2003c, 10). These vessels appeared as local Scandinavian vessels in Selling's publication. However, recent studies on Scandinavian Viking Age pottery in Sweden have confirmed several of Selling's interpretations.

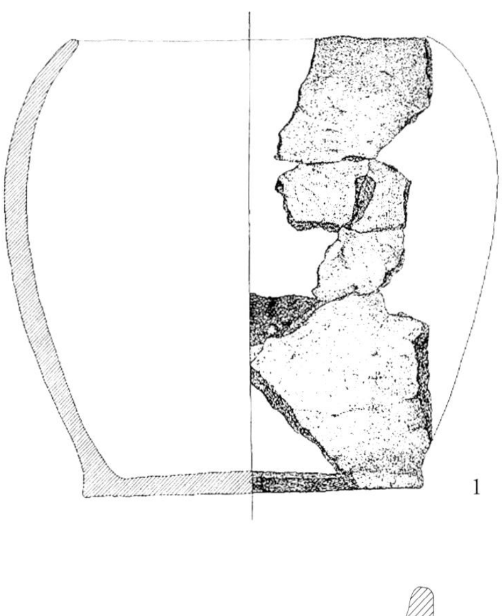

Although numerous large pottery inventories from the 8th to the 10th century have been found, only a few of them have been thoroughly investigated. One of the few exceptions is the settlement of Bjärred, outside Malmö in western Scania. More than 9 kg of pottery were documented and analysed (ID. 2002a). The typological study of the vessels and in particular the rims did not yield any significant differences between vessels from the 8th century through the 10th century (ibid. 80 f.). The dominating vessel form is bowl-shaped with inverted rim (ibid. 77 Fig. 24) *(Fig. 17,1)*. This rim type corresponds to Kempke's types Nos. 35 and 37 and represents about 75 % of the total material at Bjärred. The second most common vessel type is quite similar to the former but has a slightly everted rim shape, quite comparable to Kempke's type No. 22 (ibid. 78 Fig. 25). Such vessels represent about 20–25 % of the inventory. Other types appearing at the site at Bjärred are hanging vessels *(Fig. 17,2)* and small cups (ibid. 77 Fig. 26, 28). The hanging vessels were most likely used for cooking purposes, while the cups, decorated with knobs, may have been used as drinking jugs (HULTHÉN 1978, 109).

Almost the same distribution of vessel shapes appears at the settlement at Trelleborg in southern Scania (BRORSSON 1999, 55). At this site, 90 % of the rims are inverted. There is also a large Slavonic material, consisting of Feldberg and some Fresendorf ceramics. The pottery material is dated to the 8th and early 9th century. It is not associated to the later fortress Trelleborg, dated to the end of the 10th century. Most likely, the site served as a trading or market place rather than a settlement during the Early Viking Age.

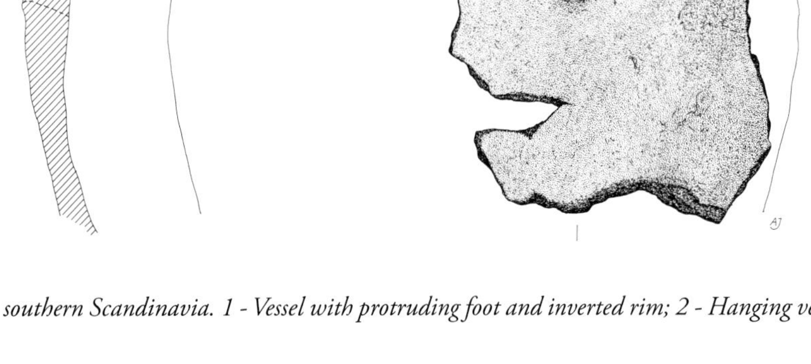

Fig. 17. *Pottery of the most common types in southern Scandinavia. 1 - Vessel with protruding foot and inverted rim; 2 - Hanging vessel. Scale 1:4 (BRORSSON 2002a, Fig. 24, 26).*

The combination of the different Viking Age Scandinavian vessel types has been confirmed by studies of material from other sites, e.g., Löddeköpinge (Ohlsson 1976, 125 Fig. 85; Brorsson 2000, 206 Fig. 13) and Järrestad (Brorsson 2002b, 350 f.) in Scania.

The vessels from former western and eastern Denmark can primarily be separated by the shape of the base. The vessels from eastern Denmark (e.g., Scania) normally had a protruding foot, while the pots from the western part (e.g., Jutland) had a rounded base.

Hans Jørgens Madsen's (1991) article concerning the pottery from Denmark is an important study of the different Viking Age vessels appearing in the region. Madsen discusses several pottery types but has a focus on what he called the "local" ones. Based on shape and decoration, these are almost identical to those from Birka and Scania. To sum up, the Scandinavian pottery can be described as consisting of a few characteristic vessel types, of which one is dominating. It is difficult to differentiate between the undecorated vessels from the Early or Late Viking Age.

These Viking Age ceramics have been found at several places outside Scandinavia. The sites are mostly in the region close to the Baltic Sea coastline, e.g., Wolin (Filipowiak / Gundlach 1992, 111) or Grobina (Nerman 1958).

Ladoga pottery *(Plate 26,3)*

The dominating vessel type in the region between Staraja Ladoga and Novgorod in western Russia during the Viking Age were Ladoga vessels *(Fig. 10,1)*. The region belonged to the East Slavs, and the pottery is characterised by everted rims and a distinct sharp shoulder. The pottery is quite similar to Menkendorf (Nosov 1992, 55) but the Ladoga vases are undecorated and have a much more distinct shoulder than West Slavonic pottery. These vessels were coil-built by hand and not shaped on turntables.

Recent studies have shown that Ladoga vessels appeared at Helgö (Rydh 1936, 45 Fig. 92) and Birka (Brorsson 2003c, 363 Fig. 2.5) in Sweden from the 8th to the 10th century.

Curonian pottery *(Plate 26,4)*

The most common type of vessel from the Curonian area, in present-day Latvia, was similar to Ladoga pottery. The body of the pots was undecorated and the vessels were coil-built by hand and not shaped on turntables. Further similarities concern the vessel shape: just like the Ladoga vessels, the Curonian had everted rims and distinct shoulders *(Fig. 10,2)*. One important difference is that the Curonian pots had finger imprints on the lip of the rim. The vessel type is rare outside Latvia but one example was found at Valsgärde (Odencrants 1933, 225 ff.) in Sweden.

West-European pottery

The West-European pottery material can be roughly divided into three groups: pottery from Saxony, from the North Sea region and finally from the Vorgebirge-Eifel region.

Selling (1955, 42 ff.) classified the West-European pottery found at Birka as AI, with several subgroups. Group AI comprises different types of pottery, such as Tating ware, Badorf pottery and also Kugeltopf pottery from the Middle Ages, found at Slottsfjärden Bay, close to the castle in the city of Kalmar in Sweden. Selling's study involved pottery from the 9th to the end of the 13th century.

Saxon pottery (Plate 22)

Among the pottery from Saxony, globular as well as flat-based vessels are represented (Capelle 1998, 109 f.). The rim shapes vary from everted to inverted. Handles may have been placed at the transition between the shoulder and the neck or close to the rim (Laux 1997, Taf. 8, 30–33). The pottery ware is rather coarse, consisting of granite-tempered fine clay. The Early Saxon pottery is mainly undecorated, but some vessels have impressions on the shoulder.

Saxon pottery can easily be mistaken for either the Slavonic Sukow pottery or Scandinavian ceramics. Based on its somewhat coarser ware and some differences in the rim shapes, it can be distinguished from the other pottery groups. Saxon ceramics differ from Slavonic in the decoration or the shape of the transition between shoulder and neck. The Sukow pottery is more or less S-shaped with slightly rounded parts. The Saxon pottery, on the other hand, has distinct transitions between the different vessel parts.

For some of these typological elements it may be difficult to put them into pottery groups, which indicates some of the problems when dealing with sherd material.

A key study of Saxon pottery is from Hammaburg in Hamburg. According to the written sources, Hammaburg was founded in the early 9th century by the Carolingians, most likely built as a defensive construction against the Slavs. The pottery was scrutinised and compared with, among other material, Slavonic ceramics. Based on the written sources, the dating of the artefacts and on natural science analyses, the material seems to be slightly younger than that from Groß Strömkendorf (Laux 1997).

Another detailed study of Saxon pottery from the same region is from a Stone Age grave at Fischbeker Heide, just south of Hamburg. Besides a considerable amount of different types of pottery from the Megalithic period, ceramics from Early Medieval times were also recovered (ID. 2003). It mainly consisted of a coarse pottery ware, belonging to the Saxon culture. In this investigation, rims were divided into 54 different types, based on the same method that H. Steuer used on sherds from Hedeby. The material from Fischbeker Heide was dated from the first half of the 8th century to the end of the 9th century (ibid. 71).

Muschelgrus pottery (Shelly ware) (Plate 23,7–9)

Muschelgrus pottery is easily distinguished from other pottery types since, as its name indicates, it is tempered with crushed seashells. The ware is full of white shells; where the shells were leached out, voids appear instead. The phenomenon of tempering with seashells ought to have had a very special purpose, most likely connected to the vessel function. The shell-tempered pottery was hand made. The seashells would have hurt the potter's hands if he/she had tried to throw it on wheel. The typical Frisian shell-tempered pot is globular with an everted rim. Decoration is rare and appears mainly as finger impressions or stamps.

Shell-tempered pottery from Northern Europe has mainly been connected to the Frisian people. The Frisians have been considered a trading people. Shell-tempered pottery appeared within Frisian settlements, but was also produced on the British Isles. Several different production sites, especially in Southeast England, have been identified. They contain, for example, the so-called St Neots Shelly ware, Thetford types and London Shelly wares (McCarthy / Brooks 1988, 179 ff.). The production of Shelly ware in England is mainly dated from the 9th to the 12th century. However, it can be found already in the Middle Saxon period, dated between 650 and the arrival of the Vikings around 800 (Laing 2003, 76). Furthermore, Lincoln Shelly ware was found at the trading site of Birka in Sweden (Stilke 1998, 580). The so-called shell-gritted wares also were discovered in, for instance, Milton Keynes in Buckinghamshire and dated to the 1st century AD (Marney 1989, 57). It has been suggested that shell-tempered ware in Frisia was restricted to the 9th century (Steuer 1974, 115 ff.), but this has been reconsidered. According to H. Stilke, the production already started in the 8th century, but not before 770, and continued into the 10th century (Stilke 1995b, 15). It is, further on, difficult to separate the Frisian Muschelgrus ware from the shell-tempered pottery from England (ibid. 11).

On the shores of the North Sea coast in western Schleswig-Holstein Elisenhof is situated (Steuer 1979), a site where nearly 2000 potsherds were discovered. About 300 of them are tempered with seashells. This type of ceramics has, in spite of the location of the site, been interpreted as imported ware. Altogether, 30 different rim types of the shell-tempered ware appear at Elisenhof. The pottery was dated to the 9th century (ibid. 25).

A fairly large amount of Muschelgrus ware also was found at Emden in Niedersachsen (Stilke 1995a). The pottery was dated to the 9th and 10th century. Muschelgrus ware constitutes about 10 % of the total material. The remaining ceramics from that period consist of coarse ware and a few sherds of imported ware from the Rhineland.

Shell-tempered pottery has been found on several places in Northern Europe. One of the first finds, outside its core area in Schleswig-Holstein, was the discovery of a vessel in grave No. 369 at Birka (Selling 1955, 60). Analyses of the ware showed that the vessel was tempered with some kind of carbonate, perhaps fossils, and produced outside Sweden (Sundius 1955, 249; Hulthén 1984, 257). Other important sites where shell-tempered ceramics were found are Kaupang in Norway, Ribe in Denmark, Dorestad in the Netherlands, Hedeby, Alt-Lübeck, Hamburg and Scharstorf in Germany (Stilke 1995b, 15 Fig. 4; Kempke 2002, Abb. 12–14). In addition, Järrestad in Scania and Lund in southern Sweden yielded shell-tempered pottery (Brorsson 2003b, 365 f.). Besides Birka and Järrestad, the most eastern finds appear in the region near the River Elbe, close to the border between Saxony and the Slavonic region *(Fig. 18)*.

Muschelgrus pottery at Dorestad was classified as fabric H-2 (van Es / Verwers 1980, 59). Such temper can be found in a large variety of vessel types but mainly in globular vessels, which are quite similar to the Kugeltopf. Other vessel types are frying pans and neckless bowls. The shell-tempered pottery at Dorestad is dated to the 8th and 9th century.

Studies of the Muschelgrus ware at Hedeby indicated a dating to mainly the 9th century. The same applies for other sites in Schleswig-Holstein and the North-Frisian islands (Steuer 1974, 115 ff.). The earliest finds of this pottery in Ribe date from 770 to 780 (Feveile / Jensen 2000, 16).

Frisian Coarse ware (Plate 23,1–6)

Frisian pottery was not only made of shell-tempered clay. Wares made of sand-tempered and naturally coarse clays occurred as well. For instance, "Eitopf", a typically stone-gritted ware of the Merovingian and Carolingian periods, was identified at Dorestad in Frisia. The type is classified as No. HIII at Dorestad, and was most likely

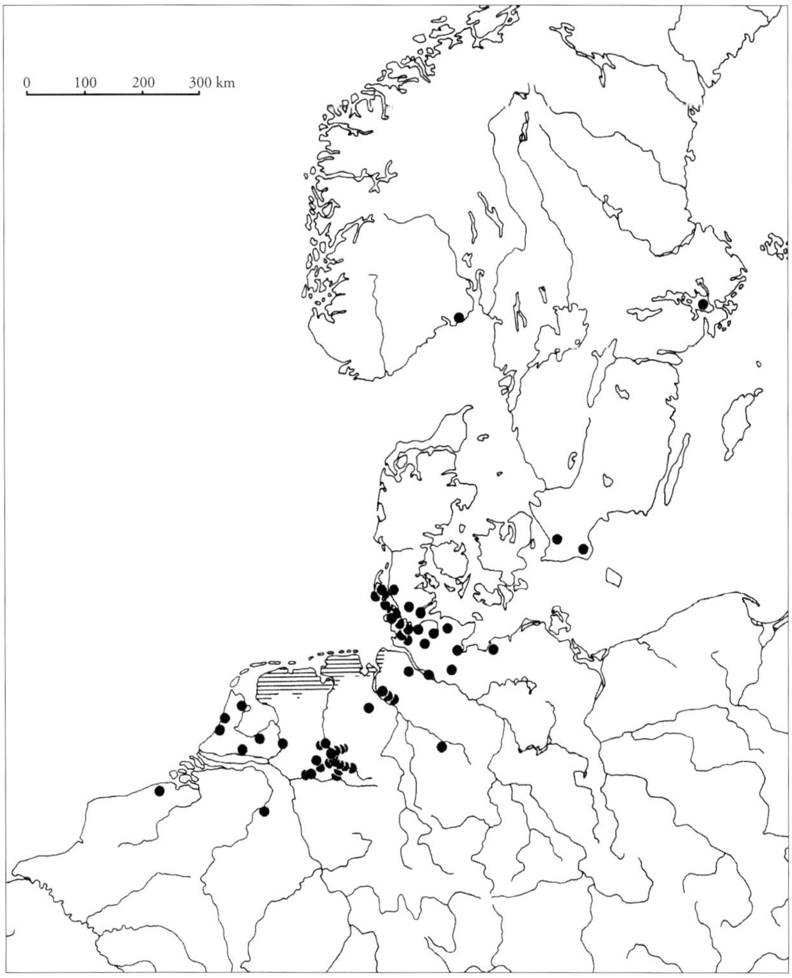

Fig. 18. Locations with finds of Muschelgrus ware in Northern Europe. Groß Strömkendorf, Lund, Järrestad and Birka have been added to the original map (after STILKE *1995a, Abb. 11).*

the successor to the "Kugeltopf" (VAN ES / VERWERS 1980, 122). The rim shape and the globular base are characteristics of the "Eitopf". This hand-made pottery without seashells but with stone grit is dated from the 8th to the 10th century.

The site of Elisenhof is situated on the borderline between the Frisians, Saxons and the Danes. There, coarse-tempered pottery dominates. A globular vessel shape with everted rim and a small flat base is most frequent. Handles also occur among the ceramics from Elisenhof. Different kinds of decoration are present, the most common being stamps or other impressions. At Emden, dated from the beginning of the 9th century to the 15th century, a large amount of "Kugeltopf" vessels of "Harte Grauware" was found (STILKE 1995a). The pottery is quite similar to the material from Elisenhof.

Vorgebirge-Eifel region

The pottery classified as originating from the Vorgebirge-Eifel area is quite easily distinguished from the other presented vessel groups at Groß Strömkendorf. Mainly the ware separates it from the Baltic Sea and the North Sea region material. The pottery from the Vorgebirge-Eifel region is wheel made, thin walled, often decorated with roulette impressions, and in some cases also glazed. During the 8th century and onwards, the traditional ware type from this region are ceramics in the Roman pottery tradition (KELLER 2004b, 155). This kind of pottery has mainly been studied in the field of art history rather than archaeology.

As mentioned above, most of the Vorgebirge pottery has been discussed and analysed outside its area of provenance, and only to a limited scale in the production district directly. The most important studies for the Baltic region cover the Dorestad (VAN ES / VERWERS 1980) and Hedeby (JANSSEN 1987) pottery. In recent years, several articles about Rhenish kilns and finds have been published. It is obvious that there is a considerable lack of knowledge about the production sites in the Vorgebirge. The large number of vessel production sites near Cologne and the similarity between them makes it difficult to separate one workshop from the other. Janssen plotted the production sites, which

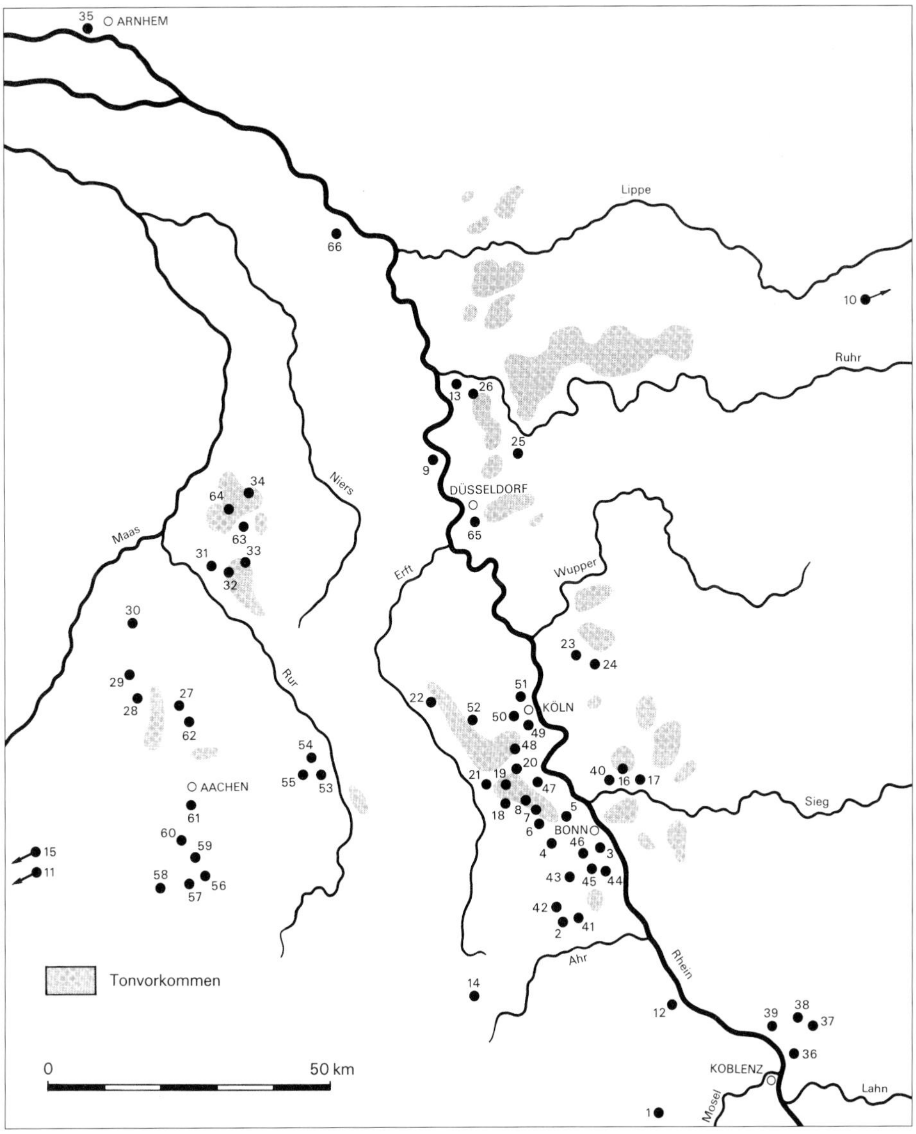

Fig. 19. *Known production sites of pottery in the Rhine and Eifel region from Early to Later Medieval times (Janssen 1987, Abb. 1). The names of the sites are listed in Appendix 5.*

were known to him when studying the Hedeby pottery. Altogether, 66 sites manufacturing different kinds of ceramics were identified in the region (ibid. 76 Abb. 1) *(Fig. 19)*.

The large amount of production sites within a limited region poses several problems. The different production sites have a close resemblance concerning both vessel shape and raw material, thus making it problematic to distinguish them. It is mainly the Badorf pottery from the Viking Age and the previous centuries that dominates the North-European archaeological literature. A parallel from the 14th and 15th centuries are the Stone ware vessels from many different sites in the same area, which are mostly referred to as Siegburg ware. This does not exclude that Badorf and Siegburg in fact dominated the pottery production in Northern Europe during their respective periods. Regarding their importance, the several small and largely unknown production sites are not sufficiently represented in the archaeological record. The pottery produced in the Vorgebirge area seems to have been rare in the Baltic region. It is mainly documented at the larger, more important sites. Because of the problem in identifying the exact production site, it may be preferable to classify these sherds as originating from the Cologne district rather than to make a doubtful specification, such as Badorf or Walberberg ware. The distance between these two sites is only 3 km, and it has been proposed to use the term "ware of Badorf type" instead of the term "Badorf ware" (Selling 1972, 41).

Analyses of thin sections may shed light on the origin of the wheel-turned pottery from the trading sites in the Baltic region. However, not only the vessels may have been transported. Sometimes, the same applies to the raw clay. It is not likely that it was brought from far away; however, according to historical sources from southern Sweden, one Stone ware factory bought clay from another, about 5 km away (von Friesen 1976, 13). Analyses of the fabric may in such a case indicate the "wrong" factory of origin for a vessel. This is one of many archaeological dilemmas.

Taking the problems regarding the exact provenance into consideration, the pottery from Groß Strömkendorf has been classified according to accepted methods and interpretations. The following gives a short presentation of the most common vessel types from the region and the types identified at Groß Strömkendorf.

Badorf (Plate 25,3–8)

As already mentioned, the best-known pottery production site in the Rhine region during the 9th and 10th century was the Badorf site *(Fig. 19, No. 8)*. Badorf consists of various pottery types, of which the dominating vessel shape is globular with an everted rim, a sagging base and a trefoil spout. Occasionally, roulette patterns on the shoulder occur. The vessel wall is thin and the surface colour is mainly pale yellow to grey. The pottery ware is sparsely tempered with sand, and the grain size normally measures between 0.05 and a maximum of 0.3 mm (Keller 2004a, 125).

Two main variants have been identified, entitled as younger and older Badorf pottery (Lobbedey 1968, 72). The older type is dated to the 8th century, the younger to the end of the 8th up to the 10th century. The shape of the vessels and the sherd thickness are the principal parameters used to separate the two variants.

Badorf pottery has been found over a wide area of Northern Europe (Stilke 1995a, 53 Abb. 28) *(Fig. 20)*. The research on the pottery from different trading places has, above all, been focused on the provenience of the ceramics. However, it is a fact that informative and scrutinised chronologies and typologies of Badorf pottery from its own core area, the Cologne district, are missing. The pottery found at Hedeby was classified as type 1 (Janssen 1987, 17 f.). It is yellow-white to yellow-grey, the surface is softly burnished and the ware is lightly tempered. The sherd thickness is normally around 5 mm. This pottery is the third biggest group of imported ceramics at the site representing 19 % of the total material, based on the weight.

Badorf pottery from the excavation at Hoogstraat I at Dorestad was classified as type W II A to W II E (van Es / Verwers 1980, 71 ff.). The great variety of shapes and wares makes it questionable if the pottery originates only from the production at Badorf or from several workshops within a limited area. The vessels have one attribute in common, i.e., the colour of the ceramics ware, which varies from yellow or pink to orange. The Badorf pottery from the Hoogstraat I excavation is dated from the middle of the 8th to the end of the 9th century.

The studies on Badorf pottery from the Vorgebirge sites have also been focused on the origin of the ceramics. The mineralogical composition of the ware played a vital role in research and for interpretations. Traditional methods, e.g., typology, have, to a certain extent, been of minor interest. Therefore, material from foreign sites, such as Dorestad, has had a major importance in the interpretation of the Vorgebirge pottery from the North European settlement and trading sites.

C. Keller (2004a) made a new approach to the Badorf, Walberberg and Hunnenschans types. Due to the vessel shape, he divided them into five different chronological groups, named A–E, based mainly on the change of the rims and the decoration. The different pottery groups were dated from the early 8th century to the end of the 10th century, with the youngest phase being represented by the Pingsdorf ceramics.

The relief-band amphora is one of the most characteristic vessels, manufactured mainly in the classic Badorf ware. It was most likely produced in the same area or perhaps even in the same workshops and fired in the same kilns as the common Badorf vessels. This amphora type was also made at other production sites than Badorf, e.g., at Walberberg. The relief-band amphora has plastic decorations on the shoulder and body, in separate ribbons. The type was classified as No. W I at Hoogstraat I at Dorestad (van Es / Verwers 1980). It is dated from the middle of the 8th to the first half of the 10th century.

At Hedeby, relief-band amphorae and Pingsdorf pottery form the largest part of the ceramic imports. A total of 35 % of the imported pottery from the settlement area at Hedeby is of the relief-band amphora type (Janssen 1987, Tab. 4). Based on the different sherd colours, the amphorae were classified as Type 5.1 and 5.2. The latter is yellow or sand to orange coloured, while the other has a grey surface (ibid. 21).

An important find of relief-band amphorae comes from a kiln in the Rheindorfer Burg in Walberberg (Böhner 1955/56, 379). Inside the kiln, a large quantity of sherd material and wasters of Badorf type were disclosed. The kiln was dated to the first three quarters of the 9th century. During the end of the 9th and beginning of the 10th century, relief-band amphorae were also produced at Duisburg, about 100 km north of Badorf and Pingsdorf (Kluge-Pinsker 1985, 54).

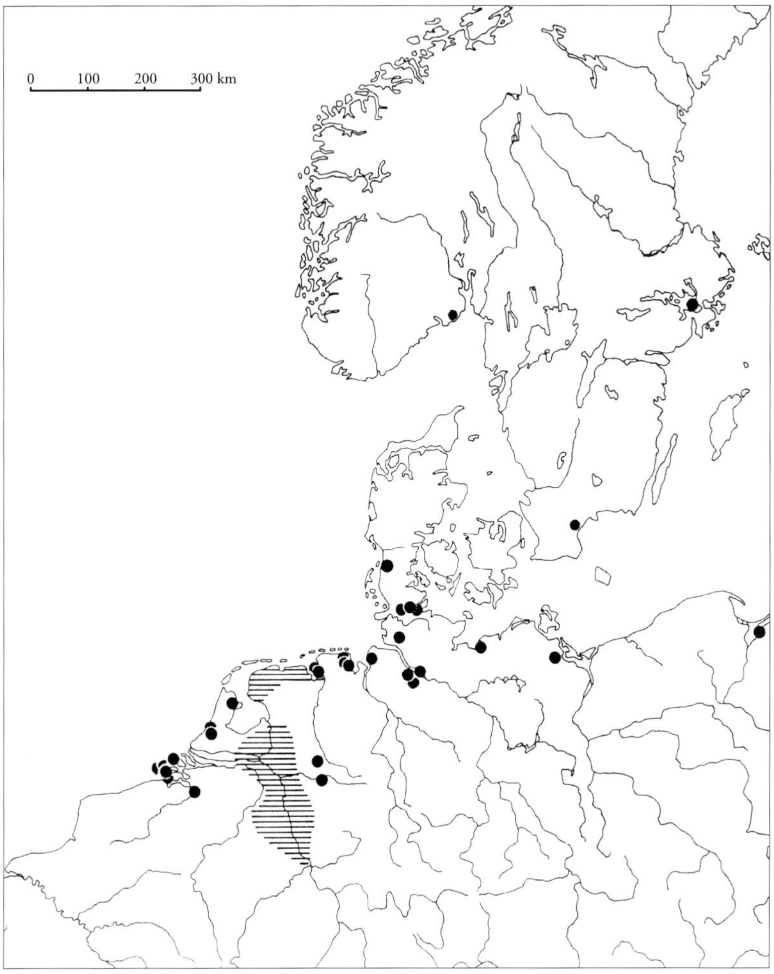

Fig. 20. Finds of Badorf sherds in Northern Europe. Åhus in southern Sweden has been added to the original map (after STILKE *1995a, Abb. 28).*

Walberberg

Walberberg was one of the major production sites of medieval wheel-thrown pottery in the Cologne district *(Fig. 19, No. 6)*. Several kilns were discovered within the city limits (BÖHNER 1955/56, 379 f.). The ceramics are similar to the "Kugeltopf" type with slightly rounded bases and everted rims. The raw material used for the Walberberg pottery mostly consists of clay, tempered with sand in grain sizes between 0.4 and 0.6 mm (KELLER 2004a, 125). Among the discoveries during an archaeological investigation at Bornheim-Walberberg in 1987 were two kilns and two pit houses. The amount of pottery material found in the kilns was very large. The ceramics were studied by C. Keller. Due to different firing atmospheres, the colour of the pottery varies: some are pink to yellow or white, some are dark grey (KELLER 1998, 299). The vessel shapes are, on the other hand, quite similar. The ware seems to be relatively homogenous but with some differences compared to the Badorf pottery.

The relation to Mayen pottery was discussed, and differences between the Walberberg and the Mayen ware were presented by KELLER (1998). The latter has a volcanic component (see below), which has not been identified in the Cologne pottery (ibid. 300).

At Walberberg, coarse-tempered pottery, produced in a Merovingian tradition, can be found in contexts up to the first half of the 8th century. The ceramics are mainly tempered with quartz sand and grog. In order to differentiate between Walberberg and other pottery wares from the region, the archaeological interpretation must rely in some cases on natural science analyses such as microscopy of thin sections.

Walberberg ceramics have been discovered at several places in Northern Europe. It was classified as type 2 at Hedeby (JANSSEN 1987, 18 f.). It is thin walled with a mainly yellow or greyish colour. At Dorestad, the Walberberg pottery was classified as variant W III A to W III F (VAN ES / VERWERS 1980, 83 ff.). It was found in layers dating from the middle of the 8th century to the end of the 9th century. It was also found at Birka (BÄCK 1995, 8).

Keller has assumed that the production most likely started in the Cologne district during the first half of the 9th century and continued until around AD 900. At that time the Pingsdorf pottery started to become the dominating type in the region (KELLER 1998, 302 f.).

Merovingian Black ware (Plate 24)

Another type of ceramics from the production sites in the Cologne district, or more accurately the Vorgebirge region, was fired in a reduced atmosphere. This firing technique resulted in a dark surface with colours from olive to dark grey (Keller 2004a, 125). Merovingian Black ware consists of vessels similar to the Badorf as well as the Walberberg pottery. Böhner (1955/56, 373) used the term "Vorgebirgsware", dating the material to the 7th and 8th centuries. It has several similarities with the Mayen pottery and was later succeeded by the Badorf ware.

A key vessel type regarding the interpretation of the Merovingian pottery is No. W VI from Dorestad (van Es / Verwers 1980, 90 ff.). It has a light-grey, bluish-grey or black outer surface. The provenience of the ceramics is uncertain; some of the regions that have been suggested are the Low Countries or the Frankish area. The pottery has been dated to the end of the 7th and beginning of the 8th century. One sherd of Merovingian Black ware was found in a pit house of the 8th century in Ystad, in southernmost Sweden[1].

Merovingian Black ware was found in great numbers in the cemeteries around the River Meuse, north and south of the Ardennes (Redknap 1998, 68). The pottery presumably originated from productions in the Frankish area that stretched over to the Rhineland in the 6th and 7th centuries. It was most likely some kind of precursor to the productions around Badorf. It is possible that two main traditions influenced the pottery assemblages in Northern Europe and England, i.e., a Frankish and a Rhenish tradition.

Mayen (Plate 25,2)

A well-known production site from the Eifel region is located in the city of Mayen, where several kilns containing large amounts of pottery were discovered (Wegner 1988, Abb. 1) *(Fig. 19, No. 1)*. The kilns were dated to two separate periods, from the 6th to the 8th century and perhaps into the 15th or 16th century. According to Wegner, the typical hard-fired material belongs to the younger phase.

The Mayen ceramics display different types and colours (brown, yellow and beige) (Redknap 1999, Farbtafel 1, 2, 3). The pottery from, for example, the 9th century consists of a fine-tempered ware of wheel-thrown pots. Typical of the pottery is the volcanic temper material (Janssen 1987, 19). The dominating vessel shape is globular with everted rim and slightly rounded base. The upper part of the pot commonly bears stamped, roulette or burnished decorations. Other Mayen vessel types are jugs, bowls and beakers. The Mayen ceramics were manufactured over several hundreds of years, influencing the colour of the ware. However, the colour variation is probably not only based on chronological differences. Numerous workshops that were contemporaneously in use made the same types of pots, which could be another explanation for the variety. At Hedeby, Mayen ceramics have mainly a reddish colour. They were classified there as type No. 3. Sherds of Mayen origin were also found at Birka (Bäck 1995, 8).

Tating (Plate 25,1)

Another distinctive production within the Eifel-Mayen region are the vessels and jugs named "Tatinger Kannen" (Redknap 1988, 15 f.). This type is characterised by a burnished surface and an often applied tin-foil decoration. The vessels have flat bases and slender bodies with curved profiles. Handles or spouts may occur. Other kinds of surfaces than the burnished black ones were also found. Especially prevalent are vases of a medium-coarse ware with surfaces of dark shades of grey. These ceramics were named after the village Tating in Schleswig-Holstein, where a well-preserved example was discovered in the early 20th century (Jankuhn 1937; Eisenschmidt 2004, Band 5.1, 171). Due to the finds of complete vessels at Birka, the name "Birka Kannen" has also been frequently used. Other names for this vessel type are "Frisian jugs" or "Tating ware".

Sherds of Tating vessels have been used by several different scholars to study and interpret the contacts in Northern Europe during the Early Medieval period. The distribution has often been connected to the distribution of basalt lava from Mayen (Müller-Wille 1985, Abb. 12). Tating ware has been found at several sites in northern Germany, Poland, Denmark, Sweden, and Norway (Müller 2002, 14 Abb. 2). The northernmost find is from Borg on Lofoten in Norway (Holand 2003, 203 ff.), the most eastern one from Staraja Ladoga in Russia *(Fig. 21)*.

Tating pottery in Hedeby was classified as type No. 6, and was, according to the surface colour/treatment, been divided into two different subgroups. At Hedeby there are sherds with both a black burnished as well as a greyish surface (Janssen 1987, Taf. 32). The same applies to the northernmost finds, which are the 54 sherds from the chieftain's farm at Borg on Lofoten (Holand 2003, 203 ff.).

Tating pottery at Dorestad was classified as type No. W VIII (van Es / Verwers 1980, 97 ff.). The vessels at the site of Hoogstraat I are dated from the middle of the 8th to the end of the 9th century (ibid. 152 Fig. 80). Group No. W XI from Dorestad is similar to No. W VIII, but

1 The sherd was identified during a recording of the material in 2004.

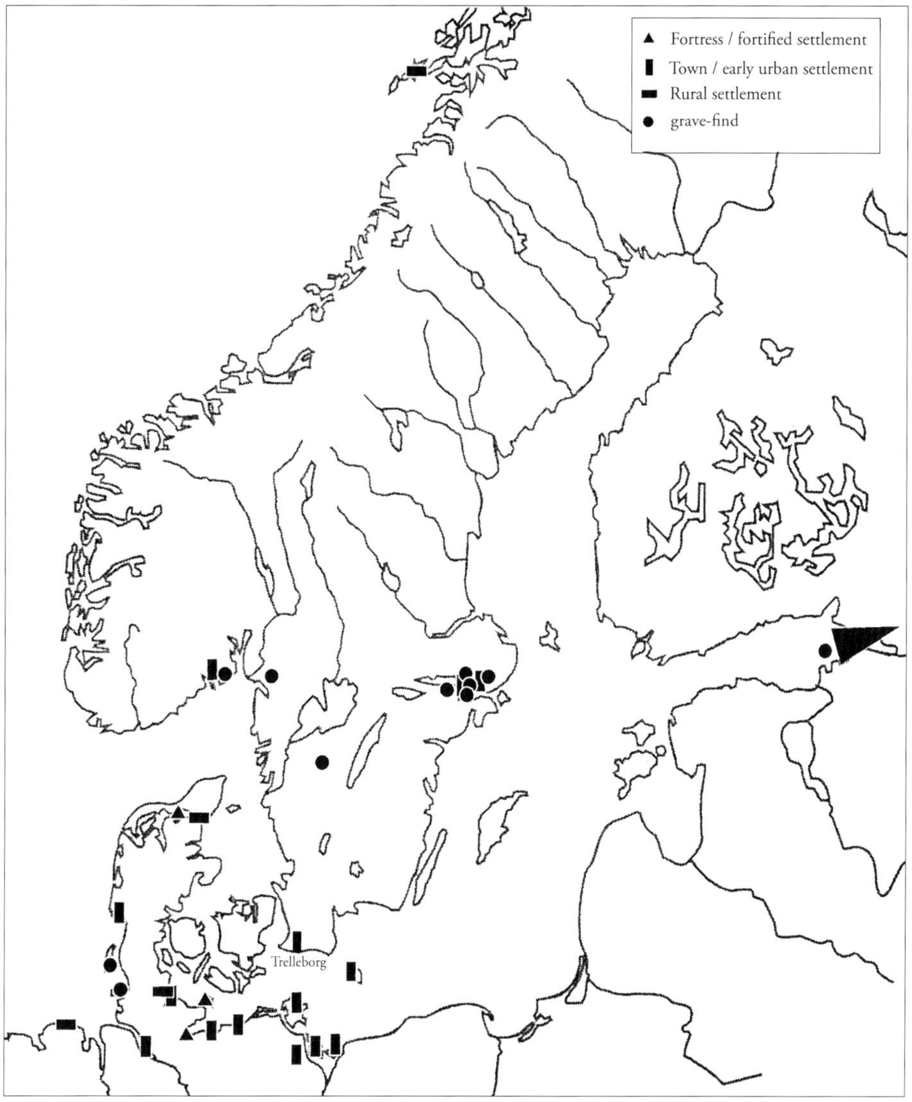

Fig. 21. Finds of Tating ceramics in Northern Europe. A recent find in Trelleborg in southern Sweden has been added to the original map (after MÜLLER *2002).*

circa 75 years older. These ceramics have either a grey or black outer surface, and the vessel shape is very similar to the Tating jugs. Tating pottery at Birka was classified as type AI:1a. According to the material from Sweden, SELLING (1955, Abb. 63) dated the ceramics to the period between 825 and 875. An example of a well-known jug type was discovered in grave No. 854 at Birka *(Fig. 22)*. This type of pottery formes only a minor part of the total material at the site. Thanks to a few nearly complete jugs, the Tating pottery has gained significantly in importance in northern ceramics research.

The most recent study of the "Tatinger Ware" was undertaken by U. MÜLLER (2002). He discussed the distribution of this pottery type in the Baltic region, its dating, and the decorated jugs with crosses, which could have had a liturgical function. Müller concluded that the different sources quite clearly show that this type of pottery was connected to religious rituals and to the early Christian mission (ibid. 18). The same interpretation was presented by other archaeologists, e.g., D. SELLING (1955, 59) and W. WINKELMANN (1984, 43 ff.). However, the hypothesis has been rejected by others, e.g., E. K. HOUGEN (1993, 27 f.) in her study of the material from the Norwegian trading site of Kaupang. The liturgical function has also been refuted by the fact that only a few jugs actually bear a cross – most of the pots were made without the Christian symbol. It has been suggested that the Tating jugs were used as objects in the wine trade (RING / WIECZOREK 1979, 360). In addition to the question about the function of the Tating ware, a debate about the origin of the production sites has taken place. A petrographic study of a sample of sherds from Kaupang showed that the ceramics were most likely made in the lower Rhine region, north of Duisburg in Germany (HOUGEN 1969, 101). Contrary analyses by B. Hulthén in the early 1970s provided data that the production was

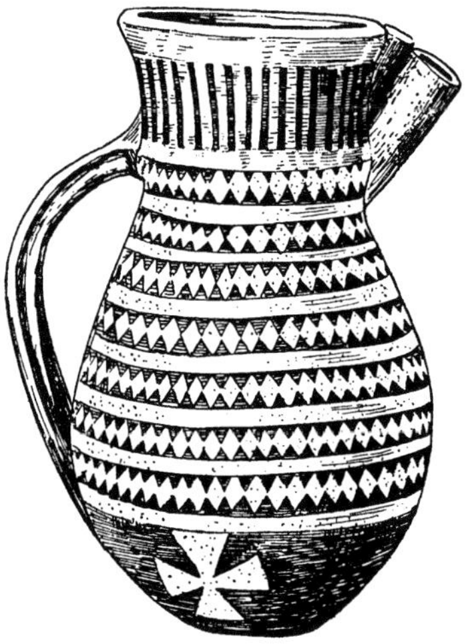

Fig. 22. Jug of Tating ware found in chamber grave No. 854 at Birka. Scale 1:4 (after Arbman 1940).

most likely connected with the monastery at Lorsch, situated in the middle of the Rhineland (Selling 1972, 41). Hulthén conducted thin section analyses of two jugs and one beaker from Birka and compared those with samples from the monastery at Lorsch. The samples consisted of local clay, a roof tile and a brick from the church wall. Hulthén's report showed that the elements in sherds from Birka support the theory that the Lorsch region may have been a production centre for the pottery in question. The clay, temper and mineralogy, especially the presence of the mineral trachyte backed this result. However, analyses performed by Richard Hodges in the 1980s on sherds from Kaupang, Ribe, Dorestad, and a number of English sites showed that at least five different groups of Tating ware existed (Hodges 1981, 65). The earlier and quite different results were, in fact, those of two different production areas of Tating ware. Hodges assumed that this special type of ceramics was not produced at a single site, but rather came from several locations. It may initially have been used for Christian ritual purposes, but unlike valuables, such as metalwork, it could be easily replicated, either with or without tin foil. During the end of the 1990s, a new attempt to find the origin of the Tating ware took place. This method involved applying neutron activation on sherds from Mayen and on Tating ware from other places to find similarities (Stilke et al. 1999; 2001). A total of 94 local sherds from Mayen and 44 sherds from Ribe, Hedeby, Kosel, Paderborn, Dorestad, and Deventer were sampled. The result was that 41 of the 44 Tating sherds formed one group and only three sherds were different. One was connected with one of the four Mayen groups. Therefore, the Hodges results, assuming several production sites of Tating ware, were rejected and it was stated that the pottery could have been made in one specific location. Considerable amounts of this type of ceramics have been found in France where they could have been produced during the end of the Merovingian period (id. 2001, 31).

The pottery from Groß Strömkendorf

The ceramic material from Groß Strömkendorf is one of the largest inventories from the Early Medieval period excavated in northern Germany. More than 62,300 sherds with a weight of nearly 780 kg were documented *(Tab. 1)*. The ceramics were found during excavations in the 1990s in settlement as well as grave contexts. The grave material represents 7.4 % of the total material, based on the number of sherds, but 15.2 % based on the weight *(Tab. 1, 3)*. This major difference can be explained by the large amount of complete vessels at the cemetery. Only about 20 % of the total material was classified into different groups. The rest was categorised as coarse hand-made ceramics. As this pottery has no decoration or other distinguishable features, it may belong to any of the vessel categories of Slavonic, Scandinavian, Saxon or Frisian origin. Of the classified pottery the Slavonic material dominates with about 90 %. Based on the weight and number of sherds, the Sukow type is the most common. It is noteworthy that of the imported vessels the Scandinavian influence is the strongest, in spite of the fact that the Muschelgrus ware and the Vorgebirge-Eifel material is much easier to identify. Several of the undefined body sherds were presumably from Scandinavian pots but they were only categorised as coarse hand-made ceramics.

Pottery type	No. of sherds	Weight (g)	% of classified vessels (No.)	% of classified vessels (weight)
Sukow	1905	104218	22.6	50.0
Feldberg	664	32306	7.9	15.5
Fresendorf	28	661	0.3	0.3
Menkendorf	98	4028	1.2	1.9
Slavonic	5005	43773	59.3	21.0
Bowl	6	51	0.1	0.0
Lid	1	16	0.0	0.0
Curonian	1	39	0.0	0.0
Ladoga	1	120	0.0	0.1
Scandinavian	182	6152	2.2	2.9
Saxon	9	2901	0.1	1.4
Frisian – Coarse ware	14	5742	0.2	2.8
Frisian – Muschelgrus ware	138	980	1.6	0.5
Badorf	47	665	0.6	0.3
Badorf – relief-band amphora	2	5	0.0	0.0
Walberberg	12	138	0.1	0.1
Mayen	48	427	0.6	0.2
Merovingian Black ware	169	3136	2.0	1.5
Tating	30	189	0.4	0.1
Ember vessel	83	3067	1.0	1.5
Classified sherds	8443	208614	100.0	100.0
Coarse hand-made ceramics	53912	570086		
Total	62355	778700		

***Tab. 1.** The studied Early Medieval ceramics from Groß Strömkendorf. The large group "Coarse hand-made ceramics" mostly consists of base sherds and shoulders without decoration or other distinguished features. The origin of these sherds is uncertain.*

The settlement

Pottery in relation to the settlement phases and dendrochronological datings

A. Tummuscheit (2003a) conducted a study of the structure of the settlement, including the spread of the pit houses and the wells at the site. Tummuscheit was able to identify a successive movement of the settlement and divided it into three phases, from the first quarter of the 8th century to the beginning of the 9th century *(Fig. 6)*. For every phase, the construction of the pit houses was analysed and described. The different phases were used as a basis for the statistical distribution of the ceramic material *(Tab. 2)*. The total amount of sherds connected to all three phases is nearly 20,000 sherds with a weight of more than 230 kg. The material used in the study consists of classified sherds, such as rim sherds and sherds of special ware (Muschelgrus ware etc.).

Area	Pit houses	Wells	No. of sherds	Weight (g)
North-eastern	15	14	1582	25454
Northern	13	12	5763	69253
Southern	24	43	3901	42303
Central	42	20	8700	94886
Total			19946	231896

***Tab. 2.** Ceramics found in pit houses and wells related to the settlements area of Groß Strömkendorf.*

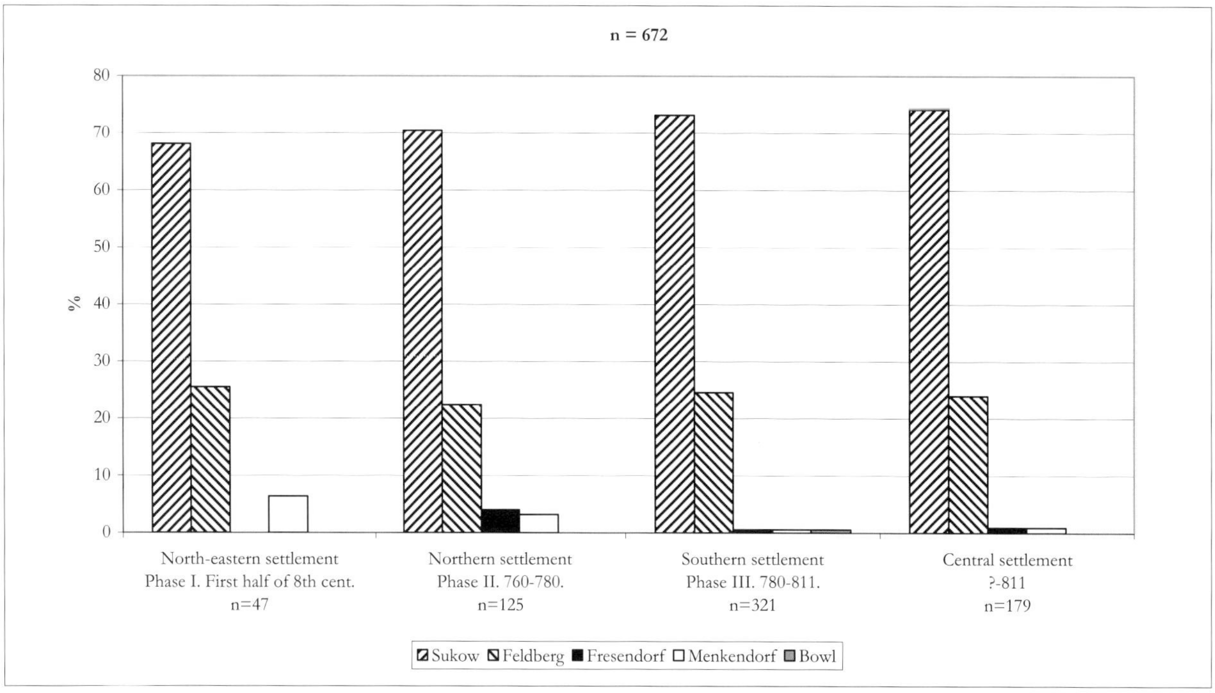

Fig. 23. *The distribution of the Slavonic pottery at Groß Strömkendorf, based on the different settlement areas and phases.*

The study of the settlement pottery involved material from different areas and the dendro-dated features which are dated to the three different phases[2]. The dendro-dated timber used for the well construction does not necessarily represent the age of the ceramics, since the sherds could have been deposited up to 30 years later *(see further p. 8)*.

Phase I: First half of the 8th century – The north-eastern settlement

The first established settlement phase at Groß Strömkendorf was dated to the first half of the 8th century. The majority of the features from this period were found in the north-eastern area, situated in the same section as the cemetery *(Fig. 5, 6)*. A major problem occurred when several features from the settlement and the cemetery were mixed together. Thus, the ceramics, as well as other types of artefacts, could have been connected to an unclear feature. This is, of course, a significant source-critical aspect of this and other investigations as well as the following archaeological interpretations. At the north-eastern settlement area, 15 pit houses and 14 wells were discovered. The dating of the cemetery is not certain but it seems to have been in use during the second half of the 8th century and the beginning of the 9th century.

The most common pottery types during the oldest phase at Groß Strömkendorf are different kinds of Slavonic pottery *(Fig. 23)*. During this phase, the Sukow pottery constitutes nearly 70 % of the Slavonic material. A quarter of the ceramics is represented by Feldberg vessels and a small amount is of Menkendorf type. The imported vessels are of Scandinavian (seven sherds), Saxon (one sherd) and Mayen (two sherds) origin *(Fig. 24)*. The material seems to be homogenous with no mixtures from other contemporaneous materials.

Altogether, 18 rim sherds were found in six different wells, dendrochronologically dated from the end of the 7th century to 736. The wells are situated in the north-eastern, central and southern areas. The sherds from the wells are classified as Sukow, Feldberg, uncertain Slavonic pots and Scandinavian types *(Fig. 25,1)*. The material is not sufficient for a reliable statistical distribution, but it indicates that Sukow was the dominating vessel type at Groß Strömkendorf during this period. The only Feldberg sherd was found close to the surface in a well in the north-eastern area. It is striking that a Scandinavian influence can be connected to the earliest phase of the settlement.

Phase II: 760 to 780 – The northern settlement

The second phase at Groß Strömkendorf was dated from 760 to 780. The settlement during this phase was concentrated in the northern area, close to the shore line *(Fig. 5, 6)*. The area was situated about 150 m south and

2 Pottery was found in the following dendro-dated features: 81-3, 152-3, 271b-3, 313-3, 327-3, 399-3, 401-3, 404-3, 473-3, 533-3, 632-3, 648-3, 676-3, 785-3, 786-3, 801-3, 804-3, 10212-17, 10290-17, 10370-17.

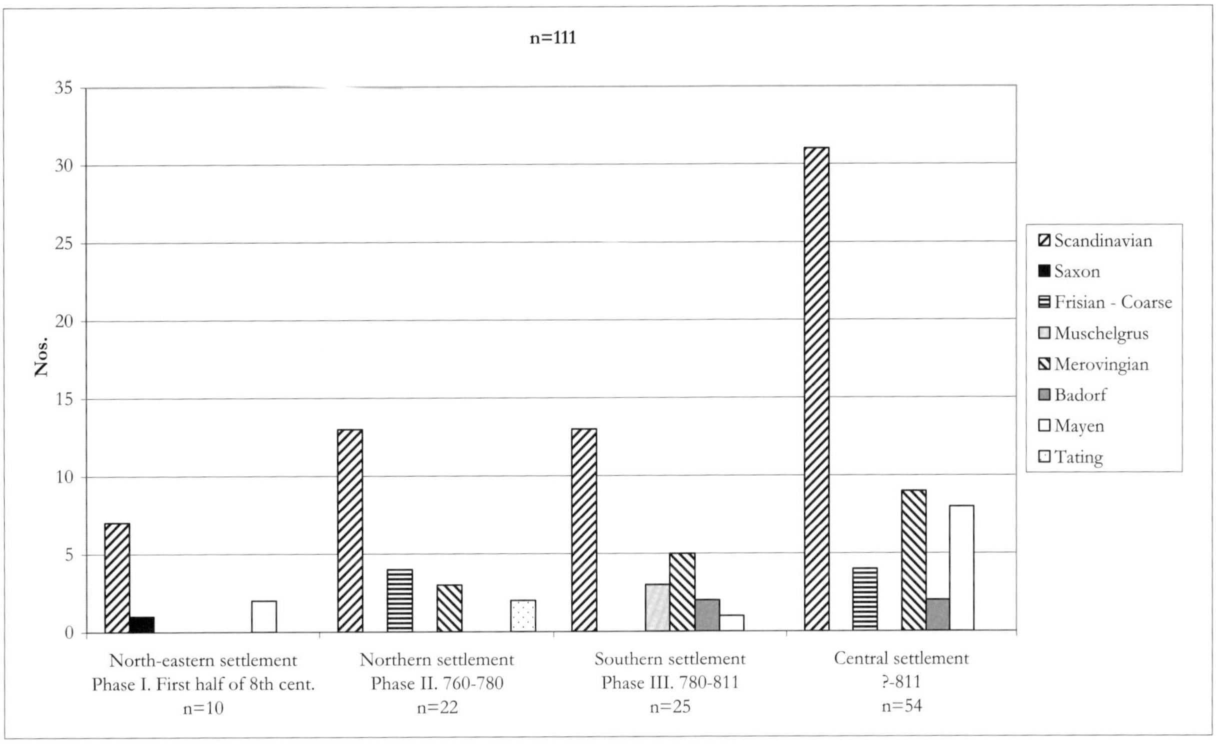

Fig. 24. The distribution of the imported pottery at Groß Strömkendorf, based on the different settlement areas.

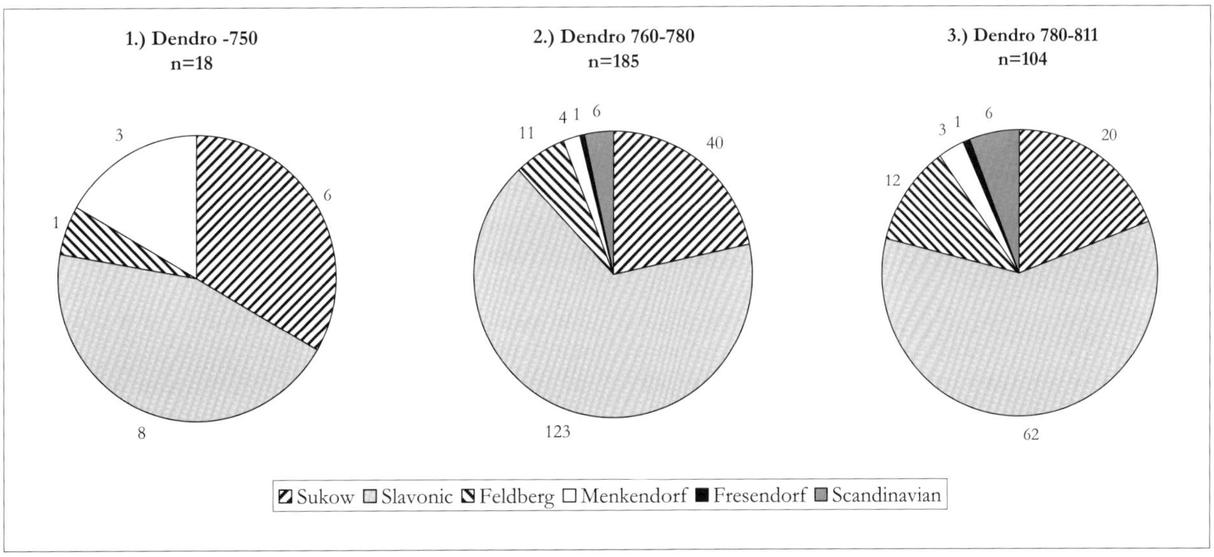

Fig. 25. The total amount of rim sherds from dendro-dated features related to the three settlement phases at Groß Strömkendorf.

100 m west of the north-eastern area. A total of 13 pit houses and 12 wells were discovered here, three wells of which were dated to 761 or 762. Slavonic pottery dominates the material with Sukow still being the most common type with slightly more than 70 % of the Slavonic rim sherds *(Fig. 23)*. The amount of Feldberg remains nearly the same as during Phase I, with slightly more than 20 % of the material. The Menkendorf pottery has decreased by a few percentage points, and Fresendorf appears as a new type at Groß Strömkendorf.

The imported vessels during the second phase are mostly of Scandinavian origin *(Fig. 24)*. Altogether, 13 sherds were identified. In addition, four rim sherds were attributed to Frisian Coarse ware, three to a Mayen origin and two sherds to Tating. A total of 185 rim sherds were found in six different wells dated by dendrochronology to between 760 and 780. The wells were situated in the northern and central area. The most common category are indeterminable Slavonic sherds *(Fig. 25,2)*. The Sukow pottery represents 22 % of the total rim material from this phase, Feldberg 6 % and Menkendorf 2 %. The oldest Fresendorf sherd was found close to the surface in a well in the central area. The only foreign vessel type recovered in the dated wells is the Scandinavian pottery with about 3 % of the material.

Phase III: 780 to 811 – The southern settlement

The section in the southern part was interpreted as the youngest settlement area at the site *(Fig. 5, 6)*. It is situated about 250 m south of the northern area. Altogether, 24 pit houses and 43 wells were found here. Nine of the wells were dated to between 780 and 811.

The most common pottery in the southern area is Slavonic, with about 93 % of the material of this origin. The assemblage of the Slavonic material is dominated by Sukow pottery: more than 70 % of the ceramics are of this type *(Fig. 23)*. Feldberg ware increased in numbers and makes up nearly 25 % of the Slavonic material. Menkendorf and Fresendorf vessel types decreased. Bowls appear at Groß Strömkendorf as a new vessel type.

The vessels of non-Slavonic origin are still dominated by Scandinavian ceramics *(Fig. 24)*. During this phase, the first sherds of Muschelgrus ware in the pit houses were identified. However, this type of pottery was also found in different pits in the northern settlement area. Five sherds of Merovingian Black ware and one sherd of Mayen origin come from the southern area. The first finds of Badorf pottery are also connected to Phase III.

Seven of the nine wells, dated to between 780 and 811, contained finds of pottery. All these wells were situated in the southern part. The distribution of the dendro-dated material does, more or less, confirm the distribution of the pottery in the southern settlement. The ceramics in the wells are mostly of indeterminable Slavonic type. Sukow represents about 19 % of the total material, Feldberg 12 % and Scandinavian nearly 6 %.

A body sherd of Merovingian Black ware was found in a well dated to 782. Two more sherds of Merovingian Black ware and one of Badorf were found in other wells, dated to between 806 and 809.

Uncertain dating – The central area

The dating of the central settlement area is uncertain. Altogether, 42 pit houses and 20 wells were discovered in the excavated area *(Fig. 5, 6)*. Three of the wells are dendro-dated to between 774 and 779. But even earlier structures were established, beginning in the first half of the 8th century. Probably some kind of activity took place during this time, but most of the settlement structures and the pottery are most likely from the third quarter of the 8th century. The character of the ceramics from the central area has much in common with the material from the northern as well as the southern part *(Fig. 23)*. The dominating pottery type is Sukow, with more than 70 % of the Slavonic material. Feldberg ware constitutes about 25 %. Other types represented in this area are Fresendorf and Menkendorf.

The imported material is dominated by more than 30 Scandinavian rim sherds *(Fig. 24)*. Other types are Frisian Coarse ware and a few sherds of Merovingian Black ware, Badorf and Mayen.

Within the central area, a large ditch (No. 266/718) was cut by several other features, being a well (No. 718d), a pit (No. 271a) and a pit house (No. 256g). The pit and the pit house were later cut by yet another well (No. 271b) and by a pit (No. 256f) respectively. The ceramic material from the different features is very similar. Pottery, such as Sukow and Feldberg, was found in the oldest as well as in the youngest features. Muschelgrus ware was recovered from the ditch and one of the wells. It seems likely that the pottery in the different features was deposited within a few years.

Conclusions about the settlement phases and the settlement areas

The distribution of the pottery sherds from the different areas in connection with the different phases has shown that the material is quite similar. The Slavonic, especially the Sukow type, is dominant in all areas. The variation in the amount between the areas is fairly low, and the only difference is that new vessel types were identified in the youngest phase and in the southern area. It seems likely that the use of the turntable and the appearance of Feldberg pottery at Groß Strömkendorf are connected with the second quarter of the 8th century and onwards. The oldest Feldberg pottery was found in a well dendro-dated to 735. The vessel still was made without the use of a turntable, making it possible that such use was introduced at Groß Strömkendorf a bit later. The Fresendorf and Menkendorf types are younger and may date from around the third quarter of the same century. The oldest dendro-dated features with Menkendorf finds were situated in the northern area. Two wells with Menkendorf pottery were dated to 761 and 762, respectively.

During all phases, Scandinavian pottery is the most common foreign vessel type. The Saxon presence was only established for the oldest phase; the Frisian ware belongs to the middle of the 8th century and onwards. The vessel parts of Vorgebirge-Eifel origin mostly belong to the same period as the Frisian ware. However, two Mayen sherds were found in settlement features in the north-eastern area. They could have been mixed with later activity at the site, but could also have been deposited during the first half of the 8th century.

The cemetery

Vessel types from the cemetery at Groß Strömkendorf

As mentioned above, the dating of the cemetery is not established for certain but it seems to have been in use during the second half of the 8th century and the beginning of the 9th century (Gerds 2006, 155). It was located in the same area as the oldest established settlement structures at Groß Strömkendorf, the north-eastern area. The total ceramic material had a weight of more than 118 kg, consisting of 4,651 sherds *(Tab. 3)*. A major source-critical problem is that 55 vessels, with a weight of nearly 46 kg, are more or less complete but were recorded as single sherds.

The most common vessels in the graves are of Slavonic origin (*Tab. 3*; *Fig. 26*). Among the Slavonic pottery, Sukow is the dominating type with about 70 % of the classified rim sherds *(Tab. 4)*. The Feldberg types comprise less than 20 %, which is the lowest quantity within the different areas of the Groß Strömkendorf site. The shares of Menkendorf and Fresendorf pottery are about 6 % and 2 % respectively. Among the grave material one sherd of an indeterminable Ladoga type was identified.

The imported ceramics from the graves weigh more than 8 kg and consist of 46 sherds and 5 more or less intact pots. The ceramics are comprised of eight different traditions or productions *(Fig. 27)*. The most common type is the Scandinavian pottery, followed by nearly the same amount of ceramics from the Vorgebirge-Eifel region. The rest was identified as Merovingian Black ware, Mayen, Tating, Badorf, and Walberberg. A few sherds and vessels are of Saxon origin, and one vessel is Coarse Frisian pottery. It is notable that no rim sherds with shell temper appear in the grave material.

Pottery in connection with different grave types

Little is known about the Early West Slavonic peoples' burial traditions. The main reason is that there is a lack of investigated burials from the Early Slavonic period. However, it has been suggested that the people used containers (such as a pot or a box) to put the ashes in. The container was placed on top of a mound, perhaps in some form of wooden shrine or house for the dead (Barford 2001; 2003). The archaeological remains of these special burials are, of course, extremely limited. Furthermore, a concentration of a pottery type could indicate a common late dating of an area of the cemetery or perhaps represent a Slavonic tribe, e.g., the Obrodites. Pottery sherds were found in every different type of grave at Groß Strömkendorf. In, for example, the chamber and coffin graves, only small pieces of ceramics were identified. The material connected to the chamber grave is an unidentified coarse sherd of uncertain origin. Rim, body and base sherds of Slavonic origin were discovered in a coffin grave, but further classification is impossible due to the fragmented condition of the sherds. However, a body sherd from Mayen was identified in the same coffin grave.

The classified ceramics, which mainly consist of rim sherds and complete profiles, were found scattered in different kinds of graves. The sherds categorised just as Slavonic were not included in the study of the Slavonic material. Imported ceramics originate from different areas, e.g., from the Vorgebirge-Eifel region.

The Slavonic vessels were mainly discovered in urn graves and in mounds *(Fig. 26)*. Other types of burials with Slavonic vessels are inhumation and boat graves. In each of these four grave types, Sukow pottery dominates. It was mostly found in urn graves and other types of cremation graves, but a fairly large amount also comes from inhumation graves. Only four pieces were connected with boat graves. Sukow pottery used in cremation graves

Grave	No. of sherds	Weight (g)
Non-classified sherds	4003	56376
Slavonic	597	53847
Imported	51	8181
Total	4651	118404

Tab. 3. *The total amount of non-classified sherds, Slavonic and imported sherds found in the graves at Groß Strömkendorf.*

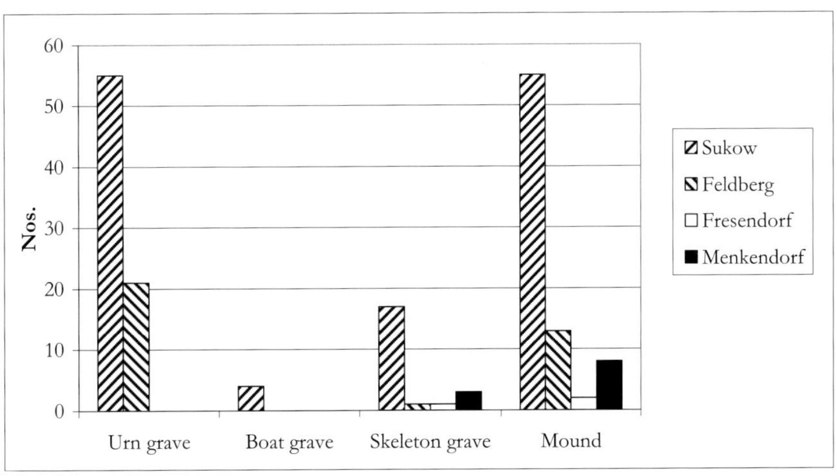

Fig. 26. *The distribution of the classified Slavonic pottery in the most common grave types at Groß Strömkendorf.*

Vessel type	No. of sherds
Sukow	131
Feldberg	35
Fresendorf	3
Menkendorf	11
Slavonic	114
Scandinavian	18
Saxon	3
Frisian	1
Merovingian Black	6
Mayen	3
Tating	2
Badorf	2
Walberberg	4
Total	333

Tab. 4. The distribution of the different Slavonic and imported vessel types in the graves at Groß Strömkendorf. The material is based on the amount of rim sherds.

was discovered in other parts of Mecklenburg, too. Some examples are Redentin in the Hansestadt Wismar, Putzar in Kr. Anklam, Todendorf in Kr. Güstrow and Rostock Dierkow in the Hansestadt Rostock (Warnke 1992, Abb. 2; Paddenberg 2000, Karte 3).

Feldberg ceramics are the second largest group of Slavonic origin from the graves. It was mainly discovered in urn graves and in mounds. A sherd with an everted rim and decoration with circles was excavated in an inhumation grave. Sukow and Feldberg pottery were found all over the cemetery, without any evidence for a clear spatial distribution.

Fresendorf as well as Menkendorf ceramics were discovered in mounds and in inhumation graves. The graves are also scattered all over the cemetery and no concentration is apparent. Altogether, 13 sherds of these pottery types were identified in the graves. However, it cannot be excluded that some sherds originally derive from the settlement in the north-eastern area.

Imported ceramics in the graves are quite limited in number, with the Scandinavian being the largest group followed the Vorgebirge-Eifel region. A small amount of Saxon and Frisian vessels was also identified.

The Scandinavian pottery is mainly connected with inhumation graves, but also appears in mounds, boat, and in urn graves. In inhumation grave No. 910a-17, a complete vessel was found *(Plate 21,1)*. Sherds from only two pots were found in boat graves. The appearance of this type of ceramics in different types of burials is not surprising. The Scandinavian influence among the urn graves could have had several different reasons. One explanation is that the type of vessel was not of any vital importance; another interpretation is that Scandinavian people had the same burial tradition as the Slavs.

The Vorgebirge-Eifel material is mostly connected to mounds and urn graves. Apart from one complete vessel in urn grave No. 800-17 *(Plate 24,1)*, only small sherds were discovered. They may have been used as *pars pro toto* in the graves. Another hypothesis is that the sherds belonged to the settlement found below the cemetery and were mixed with the grave material.

Saxon ceramics were identified in four different grave features. Three of them are urn graves and the fourth is some kind of cremation grave. The urn graves are located within the same area of the cemetery while the cremation grave is situated about 150 m south of the urns *(Fig. 28)*. The identified material consists of two sherds and two complete vessels *(Plate 22,1–2)*.

One complete but broken Frisian vessel was found in boat grave No. 206-17 *(Plate 23,1)*. It belongs to the Coarse Frisian ware. No Muschelgrus ceramics were discovered among the large amount of sherd material from the graves. This pottery was obviously not placed in the graves; it was only found at the trading site in the western area.

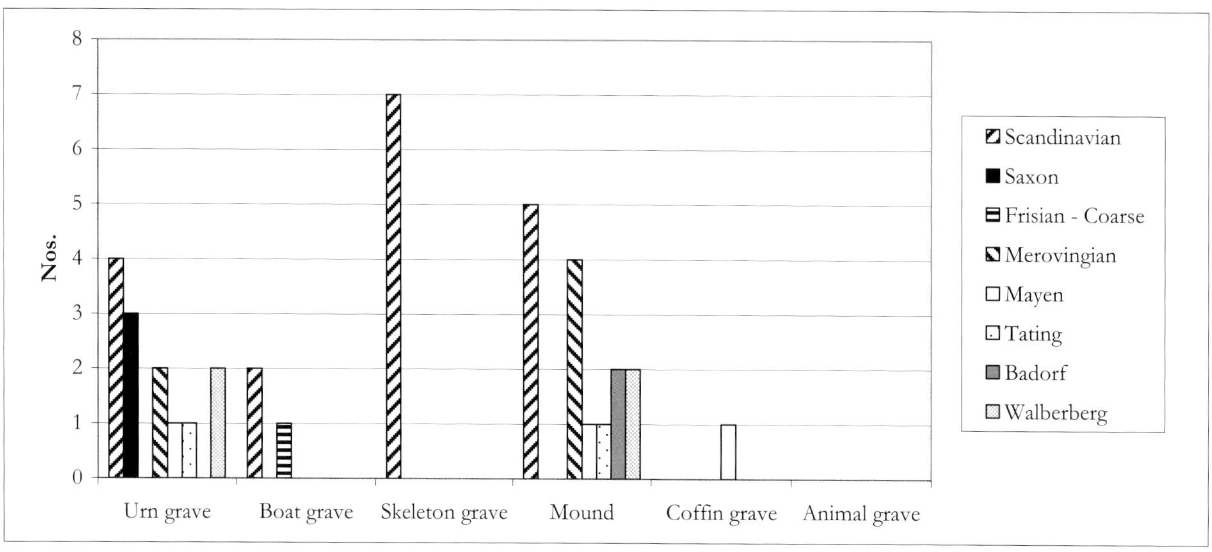

***Fig. 27.** The distribution of the imported ceramics in the grave types at Groß Strömkendorf.*

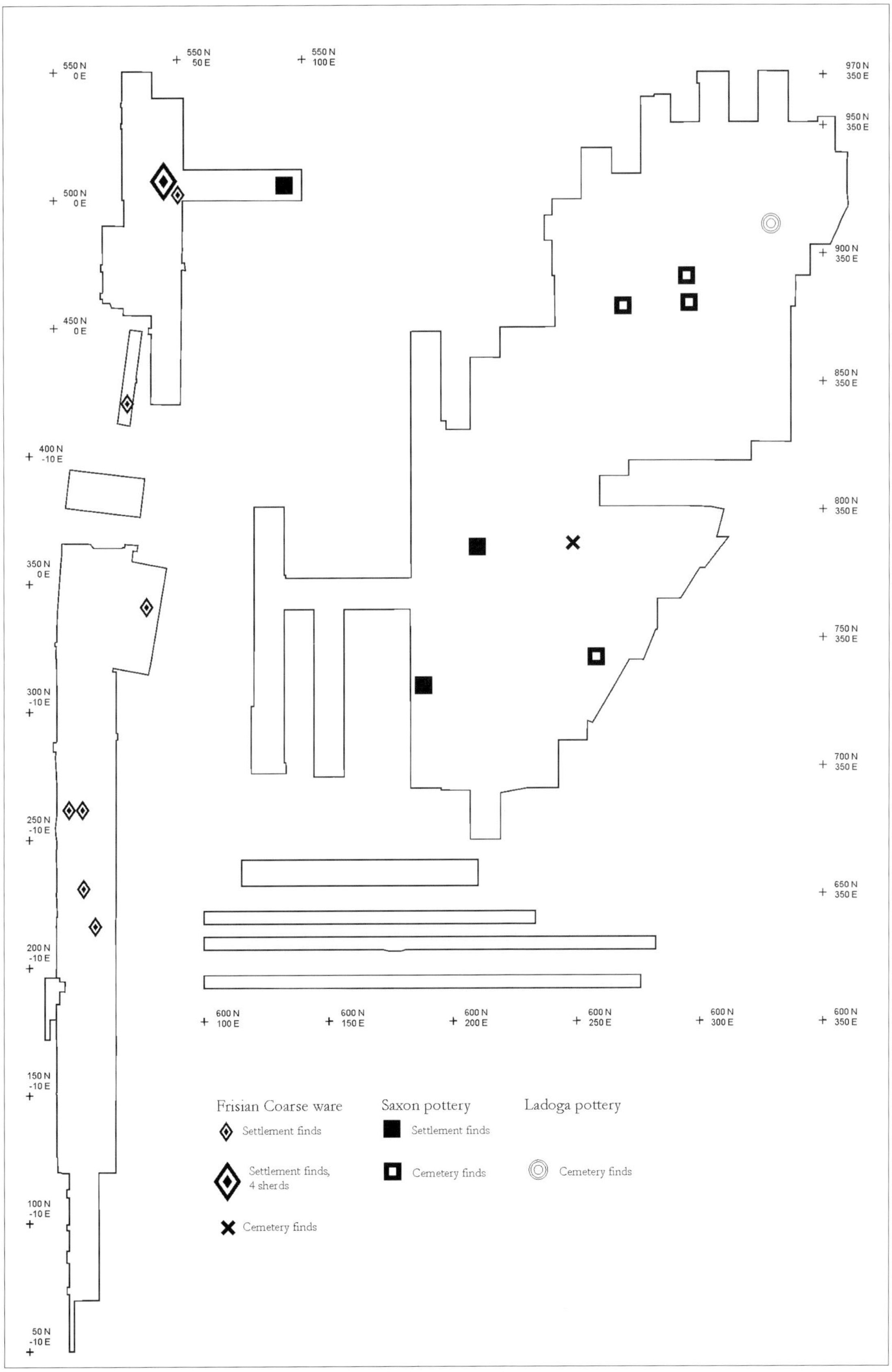

Fig. 28. The distribution of Frisian, Saxon and possible Ladoga sherds from the excavation at Groß Strömkendorf.

Conclusions about the grave material

The amount of different vessel types from the graves clearly indicates that the cemetery mainly belonged to or was used by Slavonic people (*Tab. 4*). Their ceramics were found in quite different types of burials, which could indicate rather heterogeneous burial traditions. Most of the Slavonic pottery was located in urn graves or other types of cremation graves, and it seems likely that this type of grave was connected to a Slavonic burial tradition. But it must also be stated that cremation deposits within a mound or a stone setting were a common burial tradition in, e.g., the Lake Mälaren region in the middle of Sweden at the same time period (GRÄSLUND 1980, 72).

The boat grave is a Scandinavian phenomenon; two Scandinavian vessels were found in this type of burial, as well as Sukow and a Frisian vessel.

Sukow is the most common pottery type in the inhumation graves. However, quite a number of Scandinavian sherds were also discovered in this type of burial. The Scandinavian influence in the inhumation graves is, compared to the rest of the features at Groß Strömkendorf, significant. It is, in all probability, a Scandinavian burial tradition. Several Vendel and Viking Age cemeteries with inhumation graves were discovered in Sweden. Some of the most notable are Fjälkinge (HELGESSON 1996) and Ljungbacka (SAMUELSSON 1998) in Scania and Birka (ARBMAN 1940) in the middle of Sweden. In one of the inhumation graves at Groß Strömkendorf, a Scandinavian horse fibula, dated to the 6th or early part of the 7th century, was discovered (KLEINGÄRTNER / JÖNS 2000). Unfortunately, no ceramics were found in that particular grave.

The Frisian influence in the grave material is represented by only one vessel. In connection with the lack of the Muschelgrus ceramics and other types of Frisian pots, this vessel could perhaps derive from one of the youngest burials within the cemetery.

The distribution of the different kinds of ceramics in the graves indicates that the pottery itself can hardly be used to indicate the origin of the buried person. It seems likely that different cultural groups at Groß Strömkendorf lived, more or less, close to each other and were buried in the same area. Some of the vessels may have been imported, but the Slavonic, Scandinavian and Saxon pottery could, based on the coarse quality of the ware, have been made or transported to the site by people of the same origin as the pots *(see further chapter "Groß Strömkendorf as a trading place in Northern Europe")*. The phenomenon with the Sukow vessels in the boat graves could, for example, be interpreted as a Slavonic or Scandinavian inhabitant putting such a vessel in a Scandinavian burial. This inhabitant could have lived in the same household as free person or as slave. It could, for instance, have been a Slavonic woman living in a Scandinavian household. The Sukow vessel could also have been used by Scandinavians, instead of making a new typical Scandinavian pot. The type of ceramics was probably of minor importance; rather, the structure of the burial was essential.

The Merovingian Black ware used as an urn could have been placed in the grave by a person from the Baltic Sea region. The vessel was, without any doubt, made in the Vorgebirge-Eifel region and transported to Groß Strömkendorf. The pot could, for example, have been considered as an object of status and used by a specific social group of people. It is perhaps one of the few ceramic finds that indicates a social hierarchy within the trading site.

An important parallel to the discussion about ceramics in graves comes from Valsgärde in Sweden. There, a typical Scandinavian Thor's hammer was discovered in boat grave No. 4 (ODENCRANTS 1933). The structure of the grave, the religious metal object, i.e., Thor's hammer, and the location of the grave strongly indicate a local Scandinavian burial. However, the only vessel placed in the grave was a Curonian one (ibid. 225 ff.). The vessel was most likely from the other side of the Baltic Sea (present-day Latvia) and was perhaps put in the grave by a Curonian woman. At Menzlin, for example, ship-shaped stone settings together with Scandinavian equipment were found (SCHOKNECHT 1977, 20 ff.). The only pottery in this grave was of the Feldberg type. The pottery and the burial customs seem to have been parts of a multi-ethnical exchange in the Baltic Sea region (BRATHER 1996, 163 f.). Another study of Viking Age pottery has shown that pottery is no certain evidence of the cultural affiliation of a deceased (BRORSSON 2003c).

The distribution at the trading site and the cemetery

In order to examine the distribution of different ceramic types and other attributes, e.g., decoration, size and technological aspects, distribution maps were prepared. The distribution may indicate several social aspects of the society at Groß Strömkendorf, such as where different cultural groups lived, if they were buried there and even in which part of the cemetery a burial took place. Ceramics with the same decoration may indicate contemporaneousness, which is a vital aspect of the understanding of the site and the pattern of deposition. The distribution of the ceramics may further point to diverse functions within the society.

Slavonic pottery

Large amounts of Sukow and Feldberg ceramics were found in every part of the settlement area and at the cemetery. Finds from the dendro-dated features have shown that Sukow pottery was the dominating vessel type during the whole 8th century. Feldberg pottery most likely occurred in the second quarter of the same century. Finds of Menkendorf pottery in features dated to 761/762 indicate that these ceramics appeared during Phase II, 760–780. From the third quarter of the 8th century, Fresendorf began to show. This is supported by the dendro dates, as well as by the settlement structure. The grave material, which may be dated to the same period, also consists of Sukow, Feldberg, Menkendorf and Fresendorf ceramics. Sherds from bowls were discovered at the cemetery as well as at the southern settlement area, where they were found in a well (No. 632-3) dendro-dated to 806.

As mentioned above, the finds of Sukow and Feldberg pottery are scattered all over the site. The distribution of the Menkendorf and Fresendorf pottery is presented in *Figure 29*. Two sherds of Fresendorf *(Plate 15,1.3)* were found in a pit within the north-eastern settlement. The dating of the pit and find material is uncertain and the artefacts may have been mixed with remains from the cemetery. A minor concentration of Menkendorf, as well as of Fresendorf, can be noticed in the northern part of the central area. This may be the result of the large amount of features and finds in this area.

Scandinavian pottery

Ceramics originating from Scandinavia are located all over the site at Groß Strömkendorf *(Fig. 30)*. Altogether, 179 sherds and three more or less complete vessels were found in pit houses, pits, wells, post holes, hearths, ditches, and in different grave types, e.g., inhumation graves, urn graves, mounds and in boat graves.

The distribution at the settlement illustrates that the amount of Scandinavian pottery is almost equal within the north-eastern, northern and southern parts. A higher amount is noted in the central area. Here it is concentrated in a pit house (No. 144-3) and in a well (No. 152-3), situated only a few metres from the house. In this area, a concentration of beads of the Ribe type was also found (Pöche 2005, 99 Abb. 38). Several Scandinavian sherds appeared in a pit (No. 436-3) in the same area.

The grave material is scattered over an area of 125 x 175 m. A concentration can be observed between 835 N to 880 N and 251 E to 304 E, which covers an area containing two boat graves, urn graves, inhumation graves and a mound. A single pot cannot be used to establish the cultural affiliation of the buried person. But a concentration of, for example, Scandinavian pots in one special area may indicate that this section may have been used by Scandinavian people.

Saxon pottery

The finds of Saxon ceramics are quite few in number. Only seven examples, with three of them representing more or less intact vessels, were identified at the site. The pottery was found in graves, pits and in a pit house. The finds are, with one exception, concentrated on the north-eastern settlement and the cemetery *(Fig. 28)*. The solitary sherd outside this area was found in a pit in the eastern part of the northern settlement area, about 100 m south of the north-eastern area.

No clear distribution can be observed within the north-eastern area, and the few sherds found are scattered all over this part of the site. It is obvious that the Saxon influence is connected to the cemetery and the first phase of the settlement. It may also, to some extent, be related to the second settlement phase, in the northern part.

Muschelgrus ware

The amount of ceramics with shell temper is quite substantial with a total of 138 sherds. The main reason why this group is larger than, for instance, the Scandinavian and the Saxon inventories is the fact that body sherds (95 pieces) and base sherds (4 pieces) were included in the distribution. When recording the material it is easy to identify a shell-tempered body sherd, but it is difficult to separate with the naked eye an undecorated Scandinavian body sherd from an undecorated Slavonic or Saxon one.

The shell-tempered pottery at Groß Strömkendorf is concentrated on the western part of the settlement, which is close to the waterfront *(Fig. 31)*. No finds of this ceramic type were identified at the cemetery or within the north-eastern settlement.

Within the northern part of the settlement, a concentration around 500 N 65 E can be observed. In this area, more than 20 sherds were identified. Another concentration of shell-tempered pottery is located between 205 N 10 E and 245 N 15 E, which is partly in the area between the central and southern part of the settlement. It is obvious that the shell-tempered ceramics belonged to the western part of the settlement, were not used in a funerary context and are not related with the oldest phase.

It should also be noted that these ceramics were not found in the same features or section as the Scandinavian pottery. As chronological factors are unlikely to have affected this pattern, the distribution may have had other

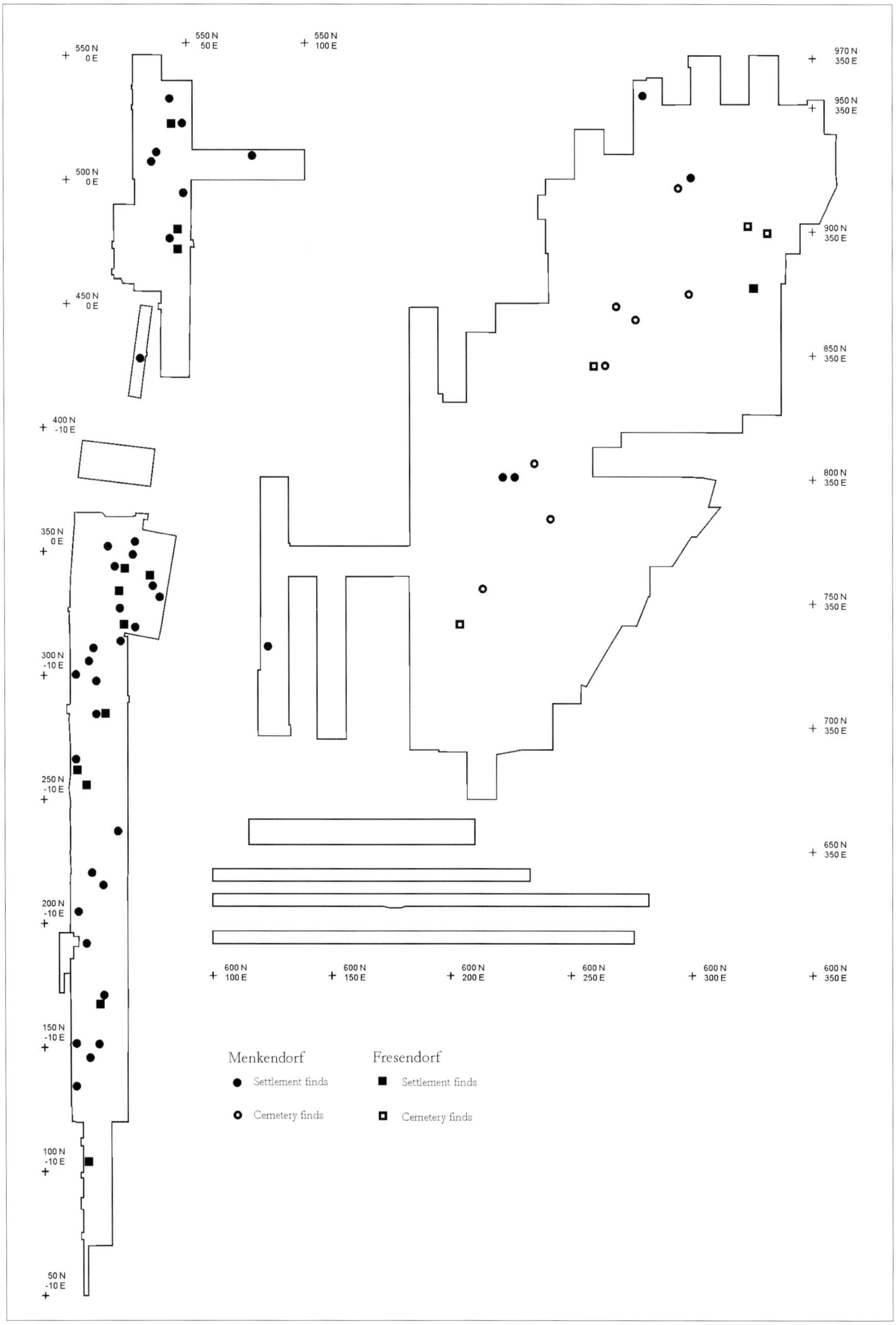

Fig. 29. The distribution of Menkendorf and Fresendorf pottery at Groß Strömkendorf.

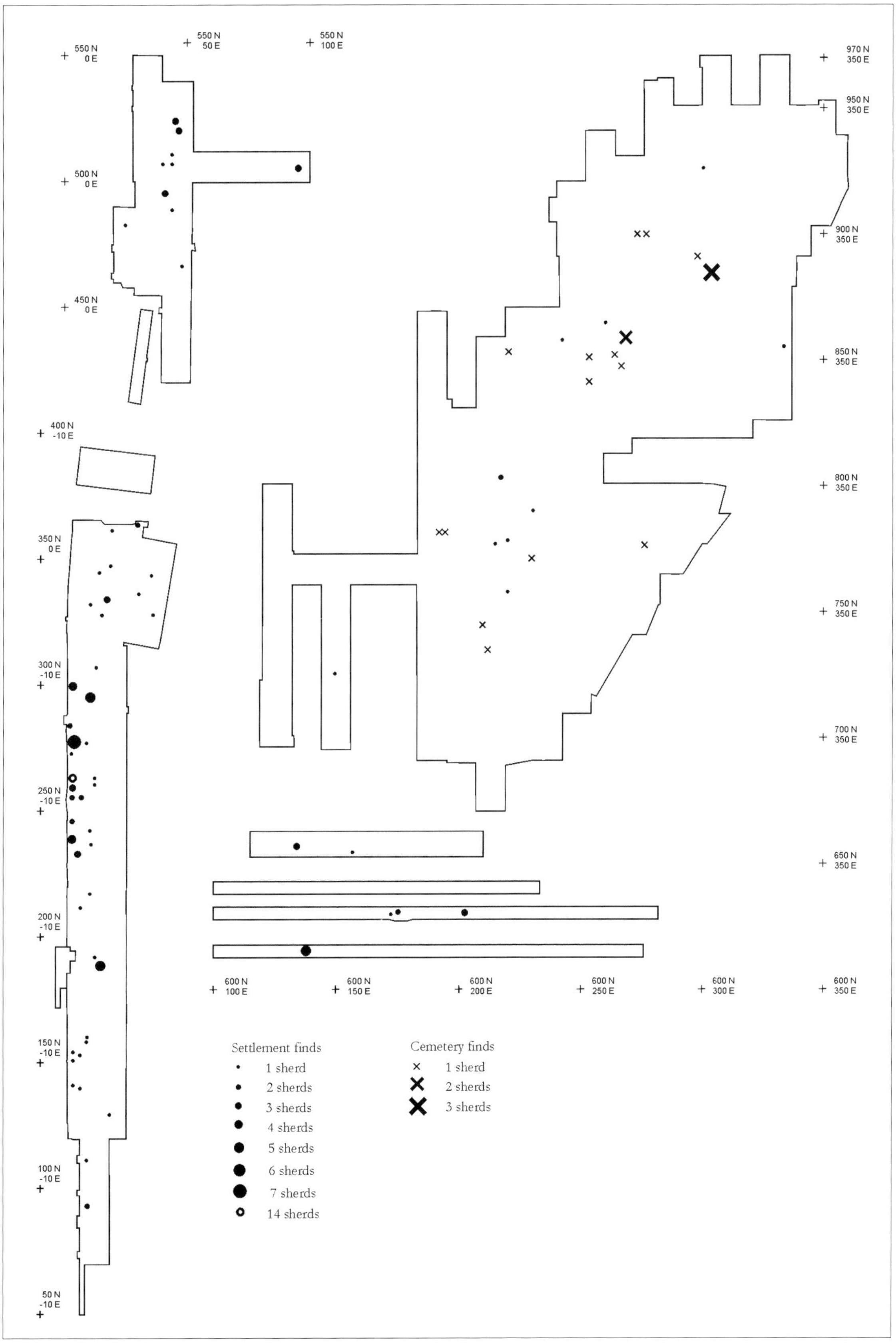

***Fig. 30.** The distribution of Scandinavian pottery at Groß Strömkendorf. A concentration of finds is noticed at the central area.*

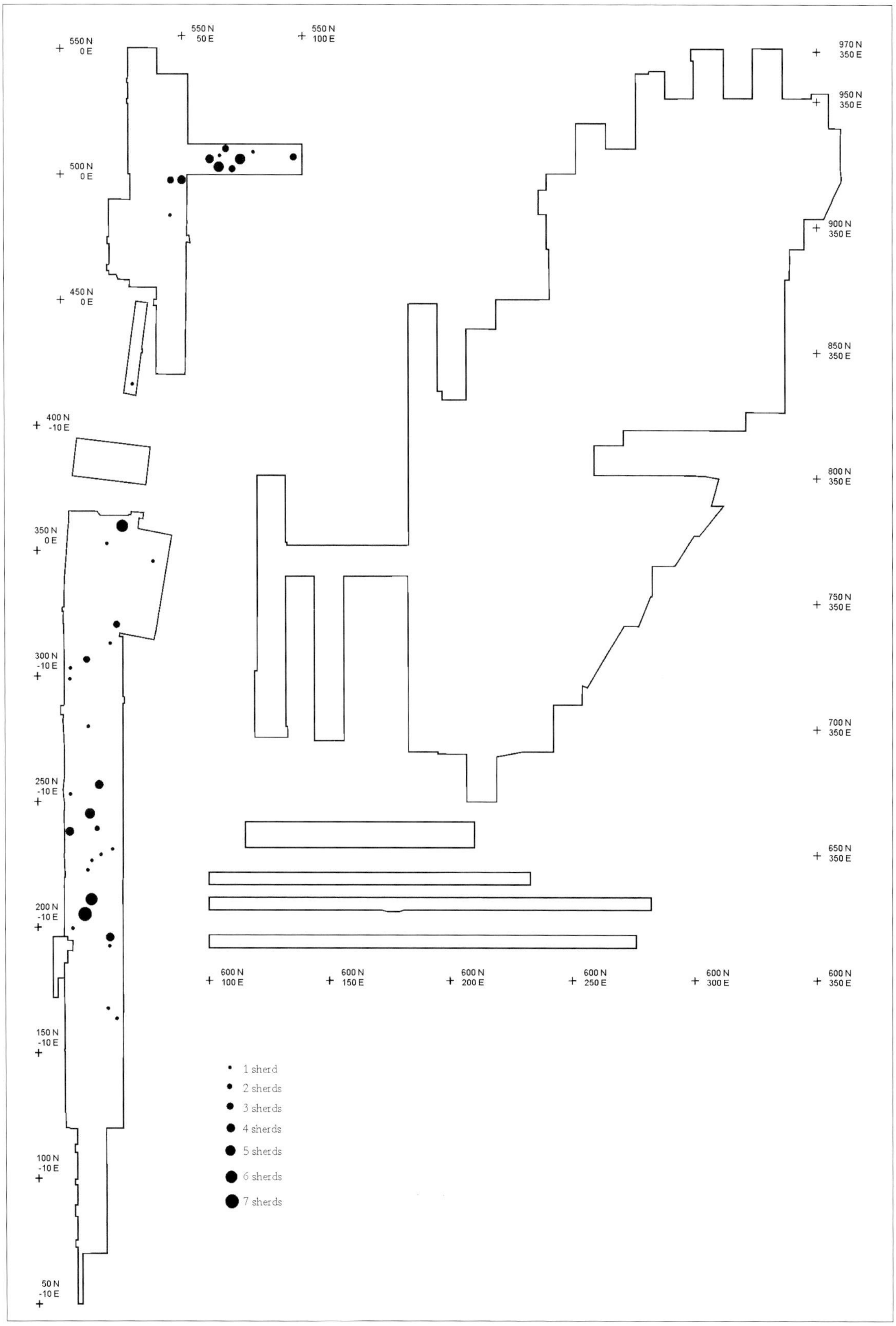

Fig. 31. The distribution of Muschelgrus ware at Groß Strömkendorf. Not a single sherd has been identified at the cemetery or the north-eastern settlement.

Vessel type	No. of sherds
Sukow	7
Feldberg	4
Menkendorf	2
Scandinavian	5
Muschelgrus	1
Frisian Coarse ware	1
Badorf – relief-band amphora	2
Badorf	1

Tab. 5. *Composition of pottery in two pits at Groß Strömkendorf (Nos. 166-3 and 238b-3) with finds of relief-band amphora.*

reasons. Perhaps different cultural groups living close to each other, but not in the same houses or not using each others' pots, have caused the distribution of the ceramics.

Frisian Coarse ware

Vessels that are most likely of Frisian origin were discovered in different features, mainly within the western part of the settlement *(Fig. 28)*. The finds are from the northern and central area. One single pot was located in a boat grave inside the cemetery area. It should be noted that the few sherds from the settlement show the same distribution pattern as the shell-tempered pottery, which supports the ethnic interpretation of the vessel type.

Vorgebirge-Eifel region

The pottery classified as originating from the Vorgebirge-Eifel region was found in several different areas at the site *(Fig. 32)*. The study involved 308 sherds of rims, bodies and bases. The reason for this is that as with the shell-tempered pottery it is possible to identity the ware type even in a body or base sherd. The most common vessel type of Vorgebirge-Eifel origin is the Merovingian Black ware, followed by the Mayen ceramics. Only a few sherds of Tating and Badorf ware are related to the settlement phases. Most of the different types of ceramics from the Vorgebirge-Eifel region were found both within the settlement area as well as at the cemetery. Three significant observations can be made based on the distribution pattern. The first is the low amount of sherds in the youngest area, i.e., the southern part. The second observation is the concentration in the southern part of the central area, near 250 N. This is especially evident in pit No. 436-3, where several sherds were located. It is the same area which contained a considerable amount of Scandinavian sherds. This pattern indicating a connection between Scandinavian and Vorgebirge-Eifel sherds is not observed within the other settlement areas or the cemetery. There is also no clear correlation between this type of pottery and any other types of foreign ceramics at Groß Strömkendorf. The third notice is the concentration of Vorgebirge-Eifel sherds in the graves in the western part of the cemetery.

Finds of two sherds of relief-band amphorae were made in two separate pits in the central area. The total classified material from both pits (Nos. 166-3 and 238b-3) consists of hand-made pottery but also of three sherds of imported (Badorf) wheel-made ceramics *(Tab. 5)*. Both pits are most likely of a late date, and the question is what the composition of the material tells us. One may suggest that social hierarchy or other aspects are reflected in the unusual relief-band amphorae. The relation to the large amount of Scandinavian sherds in the same area as the relief-band amphorae is also worth noticing.

Ember vessel

The low, flat-bottomed vessel may have been used as ember vessel, most likely as a container for charcoal or even as crucible or mould (Hulthén 1995, 20). Ember vessels are rare in the archaeological literature, but have been found, e.g., in a mound at Högom in northern Sweden (ibid. 10). The mound was dated to the Migration Period.

Finds of this type of ceramics at Groß Strömkendorf are mainly from the northern and central area of the site *(Fig. 33)*. Especially in the northern part several concentrations were found. Only one sherd is from the southern part of settlement. Sherds from a few vessels were located in the north-eastern area and at the cemetery. The interpretation of the function and distribution of these ceramics indicates that metal work took place at several parts of the Groß Strömkendorf site.

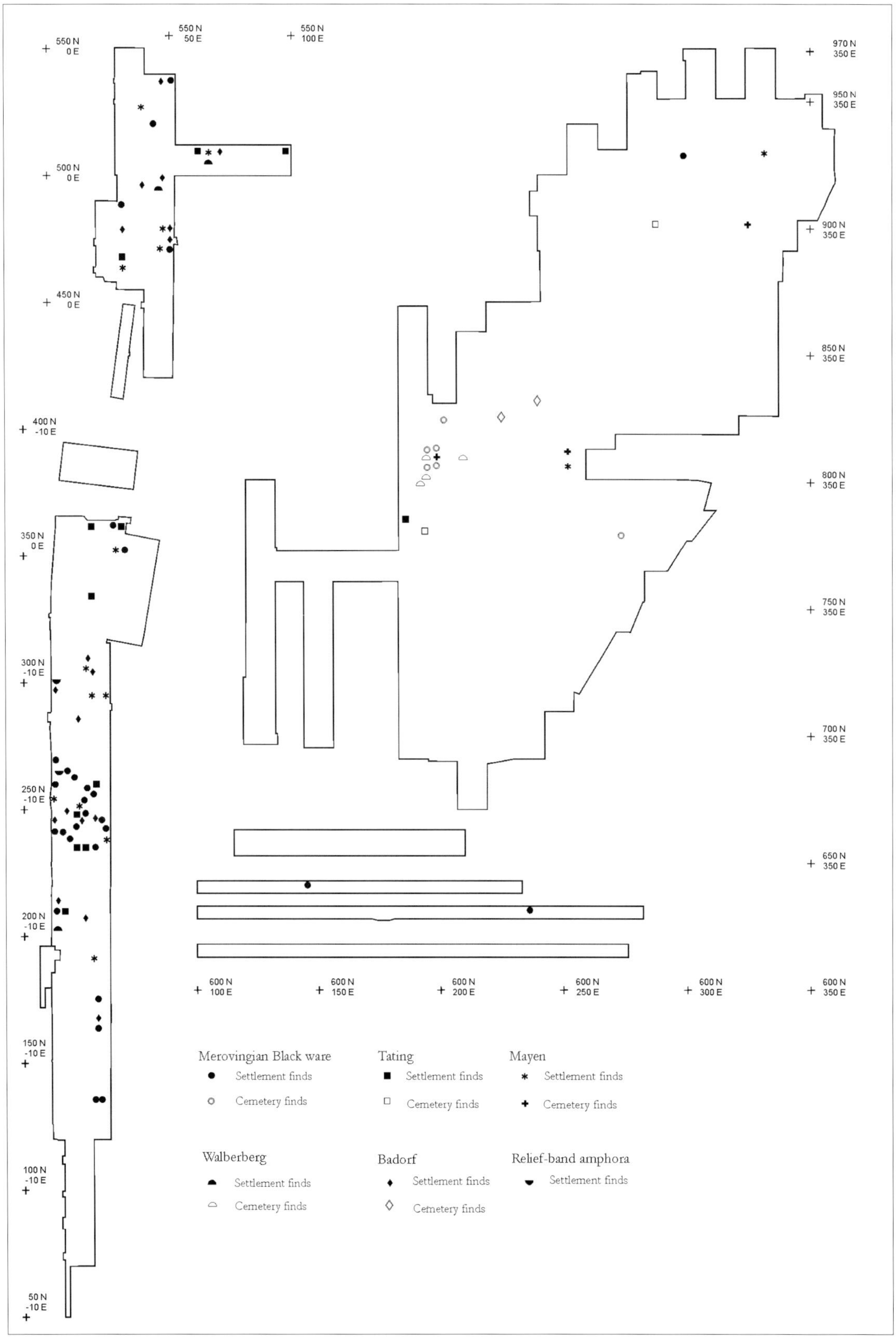

Fig. 32. The distribution of Vorgebirge-Eifel ceramics at Groß Strömkendorf. A concentration of sherds is noticed in the central area, where several Scandinavian sherds have been found as well.

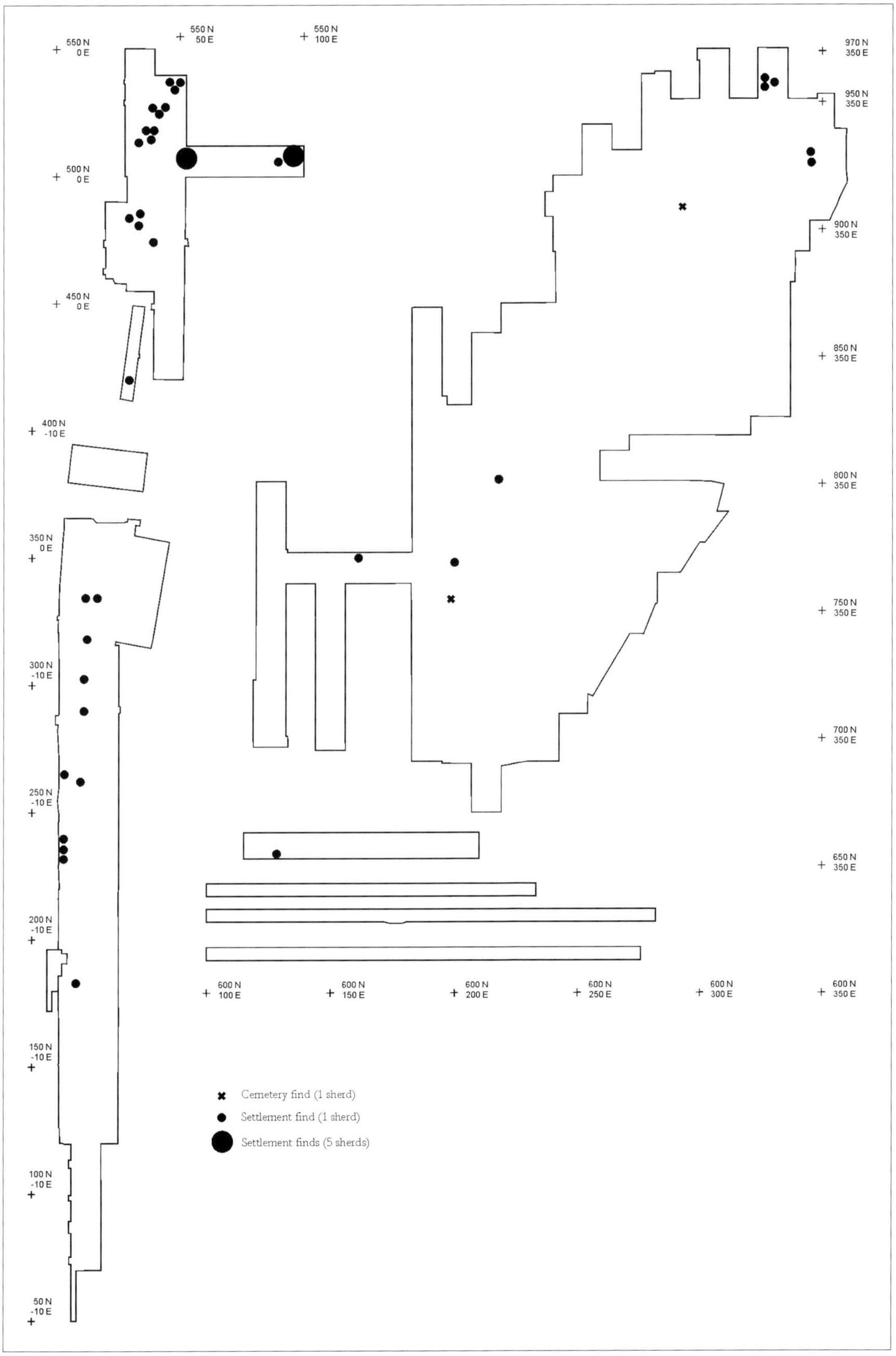

Fig. 33. *The distribution of sherds classified as ember vessels at Groß Strömkendorf. These pots were most likely used for metal work.*

Results of the study of pottery decoration, shape, and vessel size

Decorated ceramics

Of the pottery classified as Slavonic, a total of 1,981 sherds with a weight of ca. 47.3 kg are decorated. The most common decorations are wavy lines and straight horizontal lines in different compositions *(Tab. 6)*. Irregular impressions and lines in various directions are also frequently represented among the Slavonic ceramics. All these types were found all over the excavated area. Sherds with stamped decoration or cordon *(Fig. 34)* were numbered between Nos. 32 and 47, according to Meier's classification system *(Fig. 8)*. The finds of these sherds are mainly concentrated on the northern and central settlement areas, but a few sherds with stamped decoration were also identified at the other spots. The most common type is a circle in various sizes. This type was found all over the site, in graves as well as in settlement features. Decoration No. 39, a square consisting of several minor impressions, was discovered in a pit *(Plate 14,7)* in the north-eastern settlement as well as in several features in the central and southern area. In a mound, at a distance of about 15 m from the pit in the north-eastern settlement, a sherd with a flower stamp (No. 41) was discovered. Ceramics with flower stamps were also located in the central and southern area. This pattern could be the result of contemporaneous deposits. The distribution of these finds also reveals the archaeological problems involved in the interpretation of the chronology of the different features at Groß Strömkendorf.

Several types of stamp-decorated vessels are represented by only a few sherds, which is a major source-critical problem. However, the total distribution of the stamp-decorated vessels and the pots with cordons clearly indicates the following. A few sherds with this type of decoration were found in the settlement in the north-eastern area. The younger ceramics from the northern area are mostly decorated with circles of different sizes. The triangular shaped stamp from 525 N 48 E belongs to a shell-tempered sherd classified as Muschelgrus ware. A much greater variation in decoration is present in the material from the central and southern area. Here, varying types and sizes of circles, flower stamps and different cordons were found.

Ceramics with stamped decorations have been found over a wide area of the Baltic Sea region. For instance, stamps were identified at Hedeby (Hahn 1977, Taf. 16) and triangular stamps on vessel type H IA in Dorestad. This pottery type is dated to the end of the 8th century or somewhat earlier (van Es / Verwers 1980, 122). Other sites with finds of stamp-decorated pottery are Elisenhof (Steuer 1979, 138 Taf. 20), Tating Lüb (ibid. 140 Taf. 22), Kosel (Meier 1994, Taf. 12,1), Okholm (Nielsen 1998, 178 Fig. 9, 14), Bornholm (Jørgensen 1990, Pl. 19,4) and several places on Gotland (Nerman 1969, Taf. 88, 89) and in Mecklenburg (Schuldt 1956, Karte 2).

Type of decoration	No. of sherds	%
1	234	8.5
2	1032	37.5
3	15	0.5
4	849	30.9
5	103	3.7
6	16	0.6
7	1	<0.1
8	1	<0.1
10	22	0.8
12	4	0.1
13	56	2.0
14	29	1.1
15	1	<0.1
16	19	0.7
17	10	0.4
18	16	0.6
19	8	0.3
20	30	1.1
21	1	<0.1
22	1	<0.1
23	9	0.3
23a	8	0.3
24	1	<0.1
25	1	<0.1
28a	2	0.1
29	5	0.2
30	77	2.8
30a	8	0.3
31	47	1.7
31a	11	0.4
32	75	2.7
33	2	0.1
35	3	0.1
36	9	0.3
37	2	0.1
39	10	0.4
41	10	0.4
42	1	<0.1
44	2	0.1
45	1	<0.1
46	12	0.4
46a	4	0.1
47	4	0.1
Total	2752	

Tab. 6. *A total of 1981 sherds of Slavonic origin were decorated. Some of these had two or more ornaments. The classification is based on the system developed on the ceramics from Scharstorf (Meier 1990). See further figure 8.*

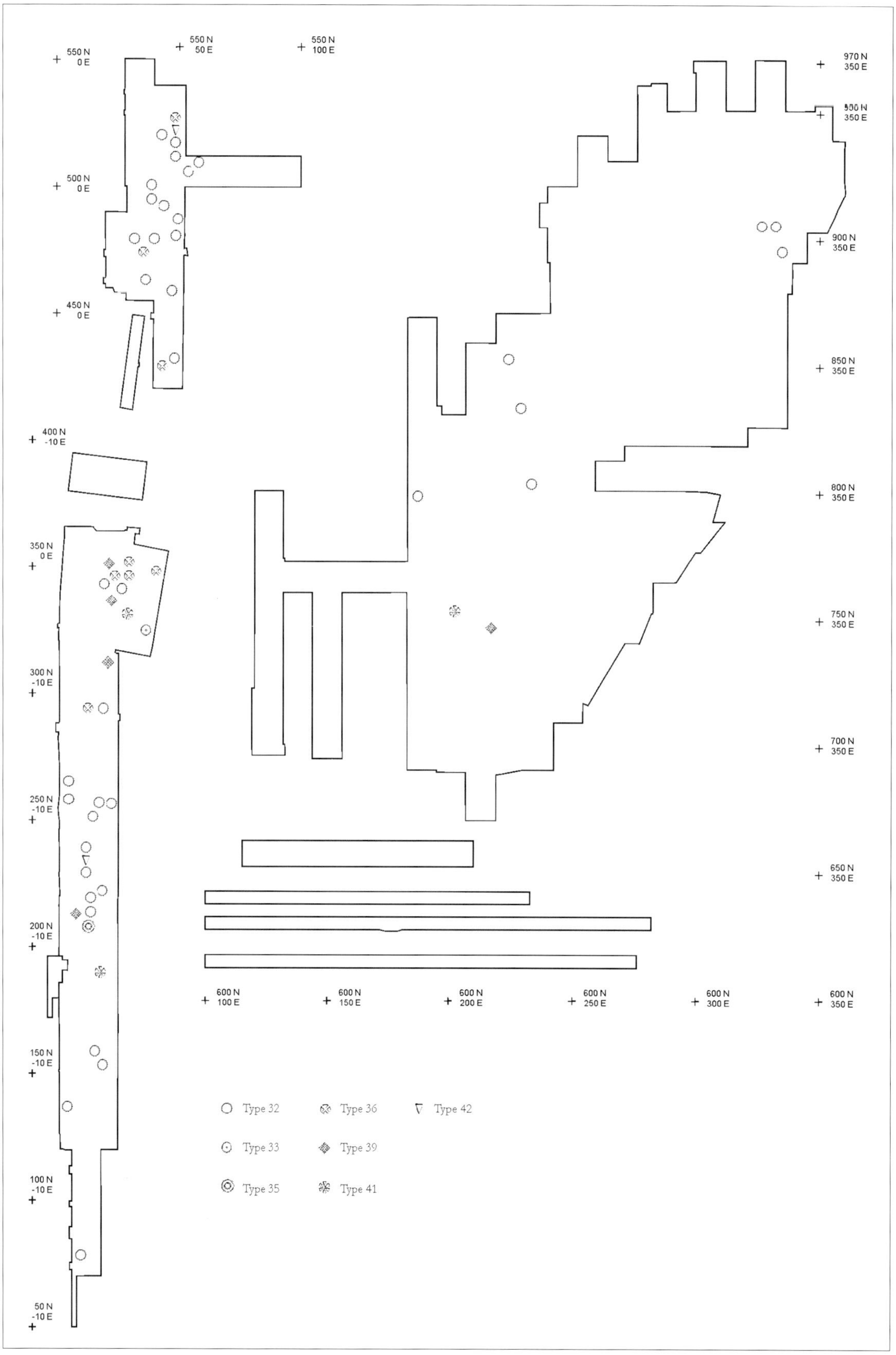

***Fig. 34.** The distribution of stamped decorated vessels according to the classification system made by* M*EIER* *(1990). The distribution is based on finds from different features at Groß Strömkendorf.*

Decoration	Sherd thickness (mm)	Maximum grain (mm)	No. of sherds	Dendro-date	Settlement
Comb impression	7.5	2.1	145	761-	Northern
Stamps (Nos. 32-35)	8.6	2.3	73	756-	North-eastern
Stamps (Nos. 36-43)	7.7	2.2	30	761-	Northern
Cordon	7.6	2.4	19	-	Central

***Tab.** 7. Macroscopic analyses of some of the younger decoration types appearing among the Slavonic ceramics at Groß Strömkendorf. The differences between these sherds and the undecorated are quite limited.*

A total of 120 different sherds or vessels are decorated with ornaments on the rim. Most of these were classified as either Feldberg or uncertain Slavonic pottery. Among the classified sherds are Fresendorf, two bowls and what appears to be a Curonian vessel. Decorated rims belong to several different types, according to Kempke's classification system. The most common decorated rim is No. 25 *(Fig. 7)*, which has a slightly everted profile and a stroked-off lip. The dominating rim type of the entire pottery assemblage from Groß Strömkendorf, No. 22, comes up with only eight decorated rim sherds.

Ceramics with rim decoration are concentrated on three different areas in the western part of the settlement. Only one single sherd is connected to the north-eastern area. Sherds or pots with decorated rim were also discovered in mounds and urn graves at the cemetery. The oldest sherd of this type from a dendro-dated context is from 762 and the youngest from 809.

Some types of decoration on the Slavonic pottery, e.g., wavy lines and horizontal lines, are often more distinct and most likely made by another method than in previous periods. The distinct ornaments may have been made when the vessel was in some kind of rotation. However, the vessels do not show any traces of being wheel-thrown and it is more likely that they were coil-built by hand and shaped and decorated on a turntable. The potter used his hands to rotate the turntable while shaping and subsequently decorating the pot *(Fig. 47,1)*.

Sherds with comb impressions were identified in every settlement area, except for the north-eastern one. There are 132 sherds with this type of decoration, of which three were discovered at the cemetery. The dendro-dated finds are from the period between 761 and 809. However, since the earliest finds belong to Phase II, is it likely that comb impressions appear on the ceramics at Groß Strömkendorf from around 760 onwards.

A Feldberg vessel with some kind of yellow or gold-like material in the wavy lines is a solitary find. Undisputably, it was applied on the ceramic surface as some kind of decoration. The substance is, in all probability, pyrite, which is a common brass-yellow mineral with a metallic lustre. The sherd was recovered as a surface find in the southern area and most likely dates to the end of the 8th or early part of the 9th century. Pyrite as decorative agent appears also on a Menkendorf sherd in Wolin. According to chemical analyses, the substance was indeed sulphur pyrite[3]. The use of such decorative elements demonstrates that people during the Early Medieval period made attempts to develop the pottery craft and to embellish the pots.

The pots with more differentiated decorations, e.g., stamps, comb impressions and cordons, were studied in detail *(Tab. 7)*. The main question was if the ware is of another quality than the common Slavonic ware. Sherds with the simple circle, Nos. 32–35, are thicker than the rest of the stamped- or comb-decorated ceramics. They are also thicker than the rest of the Slavonic pottery, including the undecorated Sukow ware. The other types of decorations (comb, cordons and stamps Nos. 36–43) are slightly thinner than the rest of the Slavonic pottery. The average maximum grain size (added or natural- coarse fractions) is about the same as for the rest of the material. To conclude, there seems to be a tendency toward thinner pots with more "sophisticated" stamps, combs and cordons than on the other Slavonic ceramics.

Grain size variation

Slavonic sherds

The maximum grain size was recorded on 5,755 rim sherds, base sherds, decorated sherds and complete vessels from the cemetery and the settlement *(Tab. 8)*. There is a large difference in quantity within the different Slavonic pottery types, both in the cemetery and the settlement. Of Sukow sherds with a mean maximum grain size of 2.4 and 2.5 mm, 1,292 come from the settlement and 167 from the cemetery. The 400 Feldberg sherds from the settlement and the 45 from the cemetery have a

3 The sherd was collected by Blazej Stanislawski, Institute of Archaeology and Ethnology, Wolin. It was analysed by the author. It contained a large amount of iron and sulphur.

	Settlement		Cemetery	
	Maximum grain (mm)	No. of sherds	Maximum grain (mm)	No. of sherds
Sukow	2.4	1292	2.5	167
Feldberg	2.3	400	2.2	45
Fresendorf	2.5	19	2.3	3
Menkendorf	2.3	55	2.2	12
Bowl	2.0	6	-	-
Slavonic	2.2	3542	2.1	214

Tab. 8. *The average grain size of the Slavonic ceramics from the settlement as well as from the cemetery at Groß Strömkendorf.*

	Maximum grain (mm)	No. of sherds
Scandinavian	2.3	173
Muschelgrus	2.0	100
Frisian	1.9	11
Saxon	2.5	8
Merovingian	0.8	44
Mayen	0.8	20
Tating	0.7	14
Badorf	0.6	24
Walberberg	0.6	6

Tab. 9. *The average grain size of the imported ceramics from all different areas at Groß Strömkendorf.*

mean grain size of 2.3 and 2.2 mm, the Menkendorf and Fresendorf ceramics between 2.2 and 2.4 mm. Six sherds of bowls were documented, with an average maximum grain size of 2 mm in diameter.

The grain size distribution connected to the different kinds of Slavonic vessels shows a homogenous ceramic craft. There is no difference in the choice of the size of the temper material between the various Early and Middle Slavonic types. Furthermore, there is no difference in grain size between the ceramics from the cemetery and the settlement. Only the bowls seem to have been made using a finer-grained temper, but we must bear in mind that only six sherds were recorded.

Imported sherds

The grain size distribution of the imported pottery varies. The Scandinavian and Saxon vessels have more or less the same maximum grain size as the majority of the Slavonic sherds *(Tab. 9)*. The Muschelgrus and Frisian ceramics seem to have been tempered with a slightly finer material. The wheel-thrown ceramics from the Vorgebirge-Eifel differ considerably from the hand-made types. All wheel-thrown vessels are made of clays with grains smaller than 1 mm.

Area	Xt (mm)	Adjusted weight (g)	Mean sherds size (cm^2)	No. of sherds
Cemetery	8.3	35.0	14.3	428
North-eastern	9.2	36.5	15.4	116
Northern	8.3	21.3	9.5	623
Central	8.2	20.7	9.1	1019
Southern	8.5	21.0	9.1	491

Tab. 10. *Calculations of the adjusted weight and the mean sherd size of the Slavonic pottery from the different areas at Groß Strömkendorf. The calculations are based on the formulas in Appendix 2. The complete or intact vessels are excluded.*

Phase	Xt (mm)	Adjusted weight (g)	Mean sherds size (cm^2)	No. of sherds
-750	14.4	21.9	8.4	29
760-780	8.2	17.0	7.6	322
780-811	8.1	14.7	6.5	167

Tab. 11. *Calculations of the adjusted weight and the mean sherd size of the Slavonic pottery from the dendro-dated phases at Groß Strömkendorf. The complete or intact vessels are excluded.*

Sherd thickness

Slavonic sherds

According to the formula presented in *Appendix 2*, the average sherd thickness was calculated on the Slavonic pottery. The average sherd thickness is a parameter based on information concerning the sherd thickness and weight of every single sherd. The sum of these values is then divided by the total weight of the material. This calculation was performed on the ceramics as a whole as well as for the different phases of the settlement and the cemetery.

The average sherd thickness indicates that the thickest pottery from the settlement is found in the oldest phase, in the north-eastern area. The thinnest settlement pottery, on the other hand, is found in the central area. The sherd material that belongs to the cemetery is about the same size as the settlement ceramics *(Tab. 10)*.

The difference in sherd thickness among the pottery from dendro-dated contexts is notable *(Tab. 11)*. There are obviously major differences between the oldest and youngest phases. The thickest pottery is found in the oldest phase with an average sherd thickness of 14.4 mm. However, the low amount of sherds in the oldest phase (29 sherds) is a source-critical aspect. Between the other two phases, the difference is only 1.7 mm (8.2 / 6.5 mm). The thinnest pottery can be found in the youngest phase at the settlement. Disregarding the small number of

	Settlement		Cemetery	
	Sherd thickness (mm)	No. of sherds	Sherd thickness (mm)	No. of sherds
Sukow	8.3	1279	8.3	164
Feldberg	7.8	394	7.9	44
Fresendorf	7.8	19	8.8	3
Menkendorf	8.0	55	7.7	12
Bowl	6.6	3	-	-
Slavonic	7.3	3409	7.3	206

Tab. 12. The average sherd thickness of the Slavonic ceramics from the settlement as well as from the cemetery at Groß Strömkendorf.

sherds and assuming that the result is representative, it may indicate that the introduction of the turntable had an effect on the thickness of the vessels.

The average maximum sherd thickness was documented on 5,588 rim sherds, decorated sherds and complete vessels *(Tab. 12)*. The measurements were, with a few exceptions, performed on the thickest part of the sherd. No measurements were made within 2 cm from the rim, on the transition between the shoulder and neck or close to the base.

The basis for the statistical evaluation of the sherd thickness varies, as there is a great variance in the number of sherds from the different Slavonic types. The vessels with the thickest walls are of Sukow type. The mean maximum sherd thickness of this type is the same at both the cemetery and the settlement, i.e., 8.3 mm. The Feldberg and Menkendorf ceramics are thinner, i.e., 7.9 and 7.7 mm respectively for the cemetery and 7.8 and 8 mm respectively for the settlement. The Fresendorf type varies from 7.8 mm at the settlement to 8.8 mm at the cemetery. But with only three sherds from the cemetery measured, the number is too small to be included in a statistical comparison. The same concerns the three sherds of bowls from the settlement. Their average sherd thickness is 6.6 mm.

A minor indication is that the mean maximum sherd thickness of the Sukow pottery is, if the Fresendorf sherds from the cemetery are not considered, 0.3–0.5 mm higher than that of the other Early and Middle Slavonic types at Groß Strömkendorf. It can be argued that this discrepancy is the result of different ways of making and building the pottery. However, the differences are so small that they also might be an accidental result. The similarities between Sukow and Fresendorf are illustrated in *Figure 35*. The distribution of the sherd thickness of the Scandinavian ceramics is similar to the Sukow and Fresendorf. The imported and wheel-made pottery of Merovingian Black ware is much thinner.

Imported sherds

The average maximum sherd thickness of the Scandinavian, Saxon and Frisian Coarse pottery is, within each pottery group, 8 mm *(Tab. 13)*. The Muschelgrus pottery is thinner; the mean value, based on 96 sherds, was calculated at 6.6 mm. The ceramics from the Vorgebirge-Eifel region are even thinner. The mean for the Badorf ware is calculated at 5.3 mm, while Tating ware is the thinnest with an average sherd thickness of only 4.2 mm.

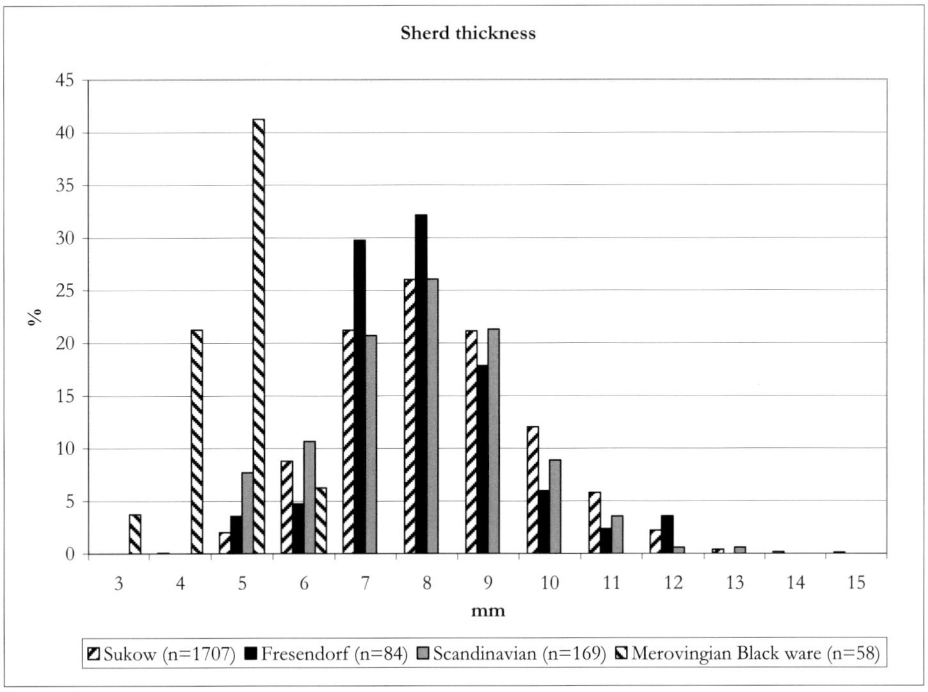

Fig. 35. The sherd thickness distribution of a sample of pottery types from Groß Strömkendorf. There are nearly no differences between Early and Middle Slavonic ceramics and the Scandinavian ware. The Merovingian Black ware from Vorgebirge-Eifel is much thinner.

	Sherd thickness (mm)	No. of sherds
Scandinavian	8.0	170
Muschelgrus	6.6	96
Frisian	8.0	10
Saxon	8.0	7
Merovingian	4.6	47
Mayen	4.7	22
Tating	4.2	16
Badorf	5.3	31
Walberberg	4.8	6

Tab. 13. The average sherd thickness of the imported ceramics from all the areas at Groß Strömkendorf.

	Settlement				Cemetery			
	Min. height (cm)	Max. height (cm)	Average (cm)	No. of sherds	Min. height (cm)	Max. height (cm)	Average (cm)	No. of sherds
Sukow	6.0	22.0	13.6	26	6.0	34.0	13.7	33
Feldberg	8.0	30.0	14.9	6	8.0	22.0	13.3	12
Fresendorf	-	-	-	-	-	-	-	-
Menkendorf	11.0	11.0	11.0	1	14.5	14.5	14.5	1
Bowl	-	-	-	-	-	-	-	-
Slavonic	15.0	15.0	15.0	1	-	-	-	-

Tab. 14. The average vessel height of the Slavonic ceramics from the settlement as well as from the cemetery at Groß Strömkendorf.

Vessel height

Slavonic sherds

The vessel height was documented on 34 Slavonic pots from the settlement and 46 pots from the cemetery. The results are presented in *Table 14*.

There seems to be nearly no difference in vessel height between the different pottery types. A larger span between small and large vessels was noticed among Sukow ceramics, but this may be connected to the fact that the amount of Sukow material is more than three times larger than that of the Feldberg ware. If one compares settlement with cemetery ceramics, the latter material is more homogenous than the pots from the settlement *(Fig. 36)*. The Sukow and Feldberg material consists mostly of pots with a height of 11 to 15 cm. With this height, the pots were suitable for, among other things, cooking purposes. Storage containers are usually considerably higher.

Imported sherds

Only eleven vessels of foreign origin can be reconstructed from the base to the rim. The average height of the Scandinavian pots is 12.5 cm *(Tab. 15)*. Two vessels of Frisian origin are 22 and 25 cm respectively and three Saxon vessels were measured to 12, 20 and 21 cm in height. The only complete pot from Vorgebirge-Eifel is a Merovingian Black ware used as an urn in grave No. 800-17. It is 21 cm in height.

The height of the Scandinavian and Saxon vessels indicates that these, just like the Slavonic pots, were primarily used as cooking vessels. The Frisian vessels are much higher and may have been more suitable as storage containers.

Rim diameter

The rim diameter of the complete vessels was also studied. The result is similar to that of the vessel height. The proportion between the height and the rim diameter can be used to project the actual height for more than just complete vessels.

Slavonic sherds

The rim diameter was measured on 402 sherds and vessels from the settlement and 76 from the cemetery. The results are presented in *Table 16*.

There is a large variation of the rim diameter within each pottery group. Small as well as large vessels were documented among the Sukow and Feldberg pottery, and it may be of interest to compare these vessels, in connection with the geographical distribution at the cemetery and the settlement. The latter shows quite interesting results.

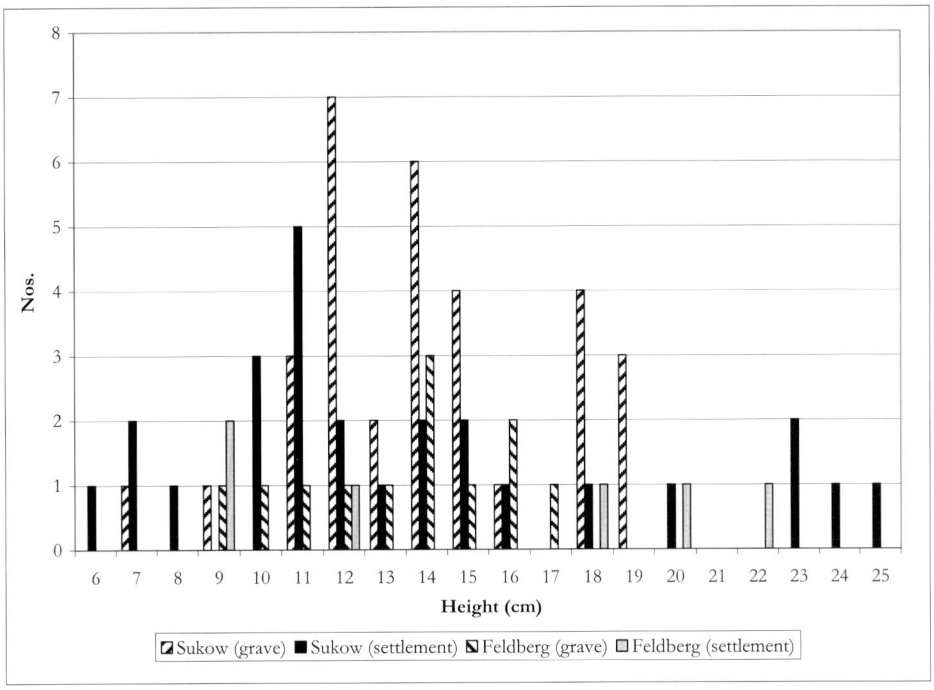

Fig. 36. The distribution of the height of the Sukow and Feldberg ceramics from the settlement and the cemetery at Groß Strömkendorf. The material from the graves is more homogenous than the pots from the settlement.

	Height (cm)	No. of sherds
Scandinavian	12.5	5
Muschelgrus	-	-
Frisian	23.5	2
Saxon	17.7	3
Merovingian	21.0	1

***Tab. 15.** The average vessel height of the imported ceramics from all the different areas at Groß Strömkendorf.*

	Settlement		Cemetery	
	Rim diameter (cm)	No. of sherds	Rim diameter (cm)	No. of sherds
Sukow	15.9	275	14.3	55
Feldberg	16.5	79	13.5	17
Fresendorf	19.0	2	-	-
Menkendorf	17.1	14	16.3	3
Bowl	-	-	-	-
Slavonic	14.4	15	22.0	1

***Tab. 16.** The mean rim diameter of the Slavonic ceramics from the settlement as well as from the cemetery at Groß Strömkend-*

The Sukow pottery from the cemetery is between 6 and 22 cm in rim diameter, but the spread at the settlement is much wider, between 6 and 34 cm *(Fig. 37)*. More than 75 % of the vessels from the graves are smaller than 16 cm in rim diameter. At the settlement, the picture is the opposite. Here, more than 75 % of the vessels are larger than 14 cm in diameter. It is obvious that most of the small Sukow vessels are from the graves, and the larger pots are from the settlement.

Almost the same pattern can be seen within the Feldberg ceramics *(Fig. 38)*. The grave material is between 8 and 22 cm in rim diameter and the vessels from the settlement are between 8 and 33 cm. The majority of the Feldberg vessels from the graves have a rim diameter less than 14 cm and the settlement pots are wider than 15 cm. The study shows that vessels meant for the graves were smaller and more homogenous in size than those for the household.

Imported sherds

The foreign hand-made vessels seem to be of the same sizes as the Slavonic ceramics *(Tab. 17)*. The Scandinavian vessels vary between 10 and 36 cm in rim diameter. The other imported pots of coarse ware have nearly the same sizes as the Scandinavian and Slavonic ceramics. The rim diameter of most of the vessels of Saxon and Frisian origin and the Muschelgrus ware are about 15–16 cm. The span within the Frisian pottery is between 11 and 27 cm. Only six sherds from the Vorgebirge-Eifel region were measured. The rim diameter of the few pots of Merovingian Black ware, Mayen and Badorf is about the same size as the hand-made pots.

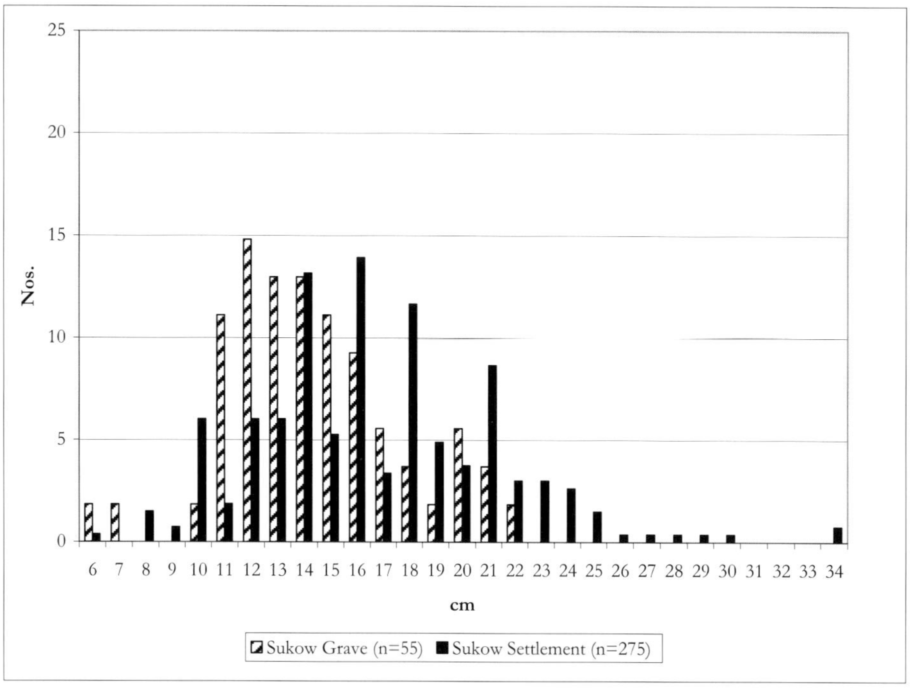

Fig. 37. The distribution of the rim diameter of the Sukow pottery from the cemetery and the settlement at Groß Strömkendorf. The pottery from the cemetery is more homogenous than the material from the settlement.

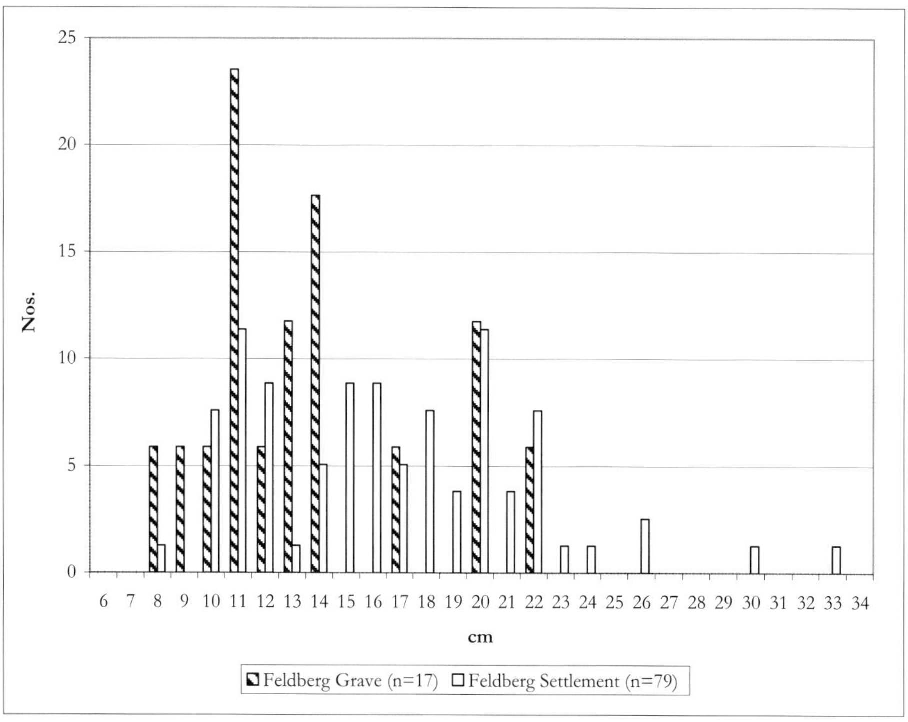

Fig. 38. The distribution of the rim diameter of the Feldberg pottery from the cemetery and the settlement at Groß Strömkendorf. The material is just like the Sukow more homogenous at the cemetery than at the settlement.

	Min. rim diameter (cm)	Max. rim diameter (cm)	Average (cm)	No. of sherds
Scandinavian	10	36	18.5	20
Muschelgrus	15	17	16.0	2
Frisian	11	27	16.5	6
Saxon	11	20	14.8	5
Merovingian	12	16	14.0	2
Mayen	16	16	16.0	1
Badorf	14	21	18.0	4

***Tab. 17.** The mean rim diameter of the imported ceramics from all the different areas at Groß Strömkendorf.*

Type of rim	Sukow	Feldberg	Menkendorf	Fresendorf	Bowl
1		19			
2		2			
4		4			
5		6			
7					
8		26	1		
9		16			
10		4			
12	1				
13		3			
14	1	4			
15		2			
16		2			
17		4			
19	7	9			
20	3	8	2		
21	69	7	3		
22	1058	115	34		
23	116	19	4		
24	162	52	3		
25	236	146	7		
26		2			
27		6		6	
28		27		11	
29	28	4	18		
30	8	3	2		
31			1		
32	2	2			
33	12	11	4	2	
34	1	4	1	3	
36				4	
37				1	
49					2
50					1
Total	1704	507	80	27	3

***Tab. 18.** The different classified rim types of the Slavonic pottery from Groß Strömkendorf.*

Rim shape

The rim shape of the hand-made pottery at Groß Strömkendorf was classified according to the typology developed by Kempke (1981a, 37 Abb. 5) *(Fig. 7)*. Altogether, 6,991 rim sherds were identified at the site, of which most could not be classified. Only 1,831 pieces were put into different Slavonic types, especially Sukow and Feldberg *(Tab. 18)*. In order to study the variation within every vessel type, the rim shape of the rim was recorded. Most of the rims belong to a few different groups. It is no surprise that odd shapes also appear among the rich ceramic material from Groß Strömkendorf.

The Sukow pottery is divided into 14 different kinds of rims. Some of the rim sherds may also be considered as belonging to other pottery types, but according to the complete profile of the vessel they were interpreted as Sukow. The same applies for a couple of Feldberg vessels. Rims classified as Sukow or Feldberg from the dendro-dated features, the different settlement areas and the cemetery, were analysed as separate units and compared with one another. According to the pottery from dendro-dated contexts, the Sukow type was divided into five different rim types during the first half of the 8th century. Everted and rounded as well as everted and stroked-off rims were identified during this period. The most common type during the first phase was Kempke's No. 22. The same applies to the next dendro-dated phase. Although the amount of different rim types increased during the last phase, the most common type was still No. 22.

To some extent the distribution of Feldberg pottery from dendro-dated contexts displays the same picture. One Feldberg sherd was found in a feature dated to the first phase. The sherd is classified as type No. 25, dated to 735. Feldberg pottery from the second phase shows some parallels with the Sukow pottery from the same period, dominated by Nos. 22 and 25. In the youngest phase, dated between 780 and 811, new rim types appeared at Groß Strömkendorf. Some of these are more or less characteristic for the Feldberg ceramics and are classified as Kempke's rim types Nos. 13 and 14.

A distribution analysis of the Sukow and Feldberg pottery in relation to the different settlement areas and the cemetery was also performed. The Sukow pottery is mainly distributed amongst five different rim types. The rims are everted with minor differences in the shape of the lip. The dominating shape for all phases is No. 22, with the highest frequency in the north-eastern area and the lowest in the southern part. The types of Sukow rims from the cemetery are about the same as from the northern settlement area.

The Feldberg pottery is also divided into five different rim types. The most common are Nos. 22 and 25, located in every excavated part of Groß Strömkendorf. The amount and distribution of the rim types varies between the different areas. It is however not possible to identify any major changes in the choice of rim types. Type No. 22, which is extremely common among Sukow vessels, is represented by at least 25 % of Feldberg vessels in every area, including the cemetery.

The frequency of the rim shapes of Sukow and Feldberg vessels points to a larger variety of shapes during the transition between the 8th and the 9th century. The most common shapes of the Sukow rims appear also on Feldberg pottery, but by the end of the 8th century or beginning of the 9th century, new Feldberg rim types appeared. This result is based mainly on the dendro dates and cannot be confirmed by the spatial distribution within the settlement phases. There is a tendency that Sukow pottery changed during the last settlement phase and that a fairly good correlation between the rim shapes of the ceramics from the northern settlement area and the cemetery exists.

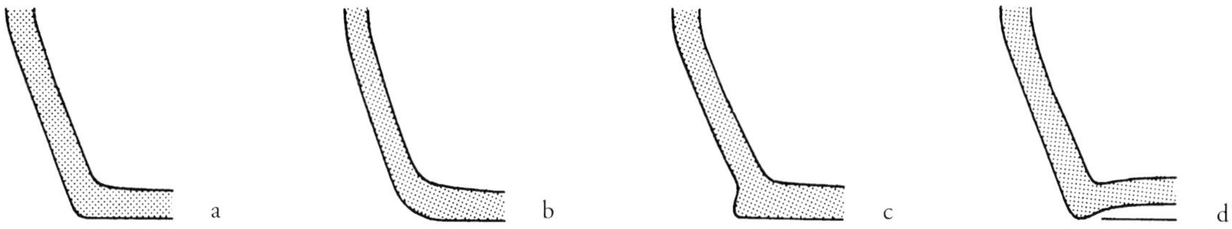

Fig. 39. *Different base types appearing on the ceramics from Kosel in Schleswig-Holstein (*Meier *1994, Abb. 15).*

The Menkendorf pots have almost the same rim shapes as the Sukow and Feldberg pottery. New shapes appear with Fresendorf and these are mostly inverted types. Stroked-off rims seem to be more common on other types of Slavonic pottery than on the Sukow vessels. The bowls have the same rims as the Garz bowls, Nos. 49 and 50.

Base types

For describing the shape of the base profiles, the classification system used for the ceramics from the 9th and 10th centuries at Kosel in Schleswig-Holstein (Meier 1994) *(Fig. 39)* was applied. At Groß Stömkendorf, the most common type is a flat base with sharp angular profile between the base and the vessel wall (type A). Most of the Slavonic pots were formed in this way. Only three sherds of uncertain origin were classified as type B, representing a rounded base. Concave bases (type D) seem to be connected with the non-Slavonic material. Altogether, ten sherds were identified. One of them is most probably a vessel from Jutland, and another is of Badorf type. The others are of uncertain origin.

The protruding foot (type C) most likely belongs to Scandinavian ceramics; this attribute is especially characteristic for the pottery from Scania. It also appears on some Slavonic vessels but not with the same distinct profile as seen on the Scandinavian material. An example of a vessel with this type of base was found in Bjärred in western Scania (Brorsson 2002a, Fig. 24) and represents a typical Scanian pot of this period *(Fig. 17,1)*. A total of 26 base sherds from Groß Strömkendorf display a protruding foot, but only one of them is classified with certainty as Scandinavian. It was discovered in pit house No. 65-3 in the central area. The other finds were scattered all over the settlement areas except for the north-eastern part. Three sherds were also discovered at the cemetery. Two wells, dated by dendrochronology to 761 and 779, contained finds of bases with protruding foot.

	Settlement		Cemetery	
	Base diameter (cm)	**No. of sherds**	**Base diameter (cm)**	**No. of sherds**
Sukow	8.5	29	8.3	36
Feldberg	8.0	6	8.0	14
Fresendorf	-	-	-	-
Menkendorf	-	-	6.7	6
Bowl	-	-	-	-
Slavonic	7.9	7	6.7	6

Tab. 19. *The mean base diameter of the Slavonic ceramics from the settlement as well as from the cemetery at Groß Strömkendorf.*

	Base diameter (cm)	No. of sherds
Scandinavian	9.9	7
Muschelgrus	9.2	5
Frisian	9.7	3
Saxon	9.0	3
Merovingian	9.0	2
Mayen	8.0	1
Badorf	10.0	1

Tab. 20. *The mean base diameter of the imported ceramics from all the different areas at Groß Strömkendorf.*

Base diameter

Slavonic sherds

The base diameter was documented for 42 Slavonic sherds from the settlement and 62 sherds from the cemetery *(Tab. 19)*. With the exception of one vessel, the diameters of the most common type, the Sukow pottery, are distributed evenly between 5 to 12 cm. The one exception is a much wider vessel with a base diameter of 17 cm. This pot was found in a well in the southern area, dated to 806. As shown in *Table 19*, the average diameters of the Sukow vessels do not differ much between the settlement areas and the cemetery. The bases of the Feldberg pottery have almost the same diameters as the Sukow. Altogether, six

	Settlement		Cemetery	
	Base thickness (mm)	No. of sherds	Base thickness (mm)	No. of sherds
Sukow	11.0	15	9.2	9
Feldberg	9.0	4	11.1	3
Fresendorf	-	-	-	-
Menkendorf	-	-	-	-
Bowl	-	-	-	-
Slavonic	9.5	10	10.1	5

Tab. 21. *The mean base thickness of the Slavonic ceramics from the settlement as well as from the cemetery at Groß Strömkendorf.*

	Base thickness (mm)	No. of sherds
Scandinavian	9.2	2
Muschelgrus	5.7	2
Frisian	6.8	1
Saxon	17.7	1
Merovingian	5.4	2
Mayen	5.9	2
Badorf	6.7	2

Tab. 22. *The mean base thickness of the imported ceramics from all the different areas at Groß Strömkendorf.*

base sherds of Menkendorf vessels from the cemetery were recorded. Their average base diameter is slightly smaller than that of Sukow and Feldberg pottery *(see Tab. 19).*

The base diameter of the Slavonic vessels indicates that the potters had a common idea concerning the size of the bases and perhaps even the size of the whole pot. The investigation shows that Sukow and Feldberg vessels usually had a quite similar base diameter.

Imported sherds

Only 22 base sherds of foreign origin were identified at Groß Strömkendorf *(Tab. 20).* The average base diameter of the Scandinavian, Saxon, Frisian and the Muschelgrus pottery are between 9 and 9.9 cm and thus slightly wider than the Slavonic vessels. Especially the Scandinavian vessels seem to have wider bases than the other types. There are very few bases that can be related to Vorgebirge-Eifel pots. Their diameters vary between 8 and 10 cm.

Base thickness

Slavonic sherds

The mean base thickness was only measured on 46 Slavonic sherds. A major problem concerning this part of the vessel is its connection to the vessel type. For that reason, 15 sherds were recorded as just being Slavonic. The results of the measurement are presented in *Table 21.* The differences between the settlement and the cemetery seem to be quite large, but it is important to bear in mind that, due to the small amount of sherds, the results are not statistically significant.

Imported sherds

The number of imported sherds, on which the base thickness was measured, is small *(Tab. 22).* Not more than two sherds of each type appear. The variation in base diameters is large, but again, the results are not statistically significant.

Impressions on the bottoms

A total of 2,315 base sherds of coarse ware were documented. It is likely that all of these vessels were built by hand and shaped with or without using the turntable. The potter could use different techniques for removing the vessels as easily as possible from the surface below. The surface such as a plate, cloth, or even sand and grass, made it also easier to rotate the vessel by hand (Lindahl / Matenga 1995, 30).

Sand in coarse fractions was noted on the bottom of 141 examined base sherds. The sand probably facilitated the turning of the vessel on the surface and its removal, compared to the raw clay surface. The same is true for the following materials.

Organic material left clear imprints under the base of a fairly large pottery group, with a total of 82 vessels bearing this kind of traces. Presumably, the vessels were placed on some kind of organic material, e.g., grass.

Textile impressions were identified on the bottom of six base sherds only. The reason for this low number may be that the remains are quite diffuse and measure only 1–2 mm in depth. It is possible that the Slavonic as well as the Scandinavian and Saxon vessels were often placed on some kind of cloth during manufacture, which left almost no traces on the vessel. The visible "imprints" on the ceramics can furthermore be used in the discussion about the quality of textiles during the Early Medieval period. They were therefore studied in detail, both macroscopically and under the microscope[4]. Two weaving types were identified. Only one vessel was, in all probability, placed on a cloth of rep weave. This vessel was found in an urn grave (No. 559). The other five vessels were built

4 The classification was made by Dr. Eva Andersson, Department of Archaeology and Ancient History, Lund University.

up or shaped on another type of cloth, the so-called tabby weave. All of these base sherds were discovered on the trading site; in pit houses (No. 65-3, 206-3), in a well (No. 156-3), in a pit (No. 256-3) and in a cultural layer (No. 488-3). The well and the pit houses from the trading site are all situated in the central area. This result, together with the fact that the only vessel made on a rep textile was found in the cemetery, is worth noting.

The size of the stitches indicates that the material used for the cloth was most likely linen. Rep as well as tabby weave were common during the Viking Age and usually used for different kinds of garments. A study of textile remains from Slavonic graves in northern Germany, dated to the Carolingian period, showed that the most common type of textile fragments were made by tabby weave. This was also the case with the Germanic (Saxon, Frisian and Scandinavian) graves from the same period (Bender Jørgensen 1992, 80 f.). Slavonic pottery with textile impressions was also found at Menzlin (Schoknecht 1977, 210 Taf. 51).

Other types of impressions identified may be a result of the pots being placed on a turntable during manufacture. The total amount of sherds with this kind of impression is fairly low, only 21 pots show any trace resulting from a turntable. It is not possible to identify a complete motif. The best examples are shown in *Plate 18*. Some sherds have circular imprints at the centre of the base. This is most likely a mark made by the central axle of the turntable. Admittedly, only one complete vessel with an imprint from a turntable was found. It is a Menkendorf vessel dated to the second half of the 8th century or the beginning of the 9th century. There is no evidence among the base sherds that the Sukow pottery at Groß Strömkendorf was made or shaped on a turntable.

Fragmentation

In order to investigate the intensity of the activities at the Groß Strömkendorf site, the adjusted weight and the mean sherd size were calculated *(App. 3, 4)*. The calculation of the adjusted weight shows three partly different numbers of sequences. The largest sherds with an adjusted weight of about 36 g can be found at the cemetery and in the north-eastern part *(Tab. 10)*. The sherds from the north-eastern area are much larger than the rest of the material from the settlement. The other areas at Groß Strömkendorf show quite similar results concerning the adjusted weight.

The ceramics from dendro-dated features show that sherds with the smallest weight were found in the structures from the youngest phase *(Tab. 11)*. The largest sherds are from the first phase.

The mean sherd size illustrates, not surprisingly, the same pattern as the adjusted weight. The largest sherds were found at the cemetery and at the north-eastern part *(Tab. 10)*. The differences between the other settlement areas are limited. According to the dendro-dated features, the sherds have about the same size in the separate phases *(Tab. 11)*. However, the smallest sherds are related to the youngest phase.

The fragmentation of the pottery from Groß Strömkendorf shows that the largest sherds are found within the cemetery and in the oldest settlement phase. The main reason why large sherds were found at the cemetery may be that no later activity was identified in this area. Since the area was a religious place, it was most likely not used for other activities when Groß Strömkendorf had a function as a trading place. The fact that sherds were only little fragmented in the north-eastern settlement area may account for a low or no later settlement activity taking place here. A source-critical aspect may be that grave material was mixed with features interpreted as settlement structures.

In the other phases of the settlement, the pottery sizes remained nearly the same, with a minor difference in that the smallest sherds were found within the youngest area. It may indicate that the activities at Groß Strömkendorf increased during the transition from the 8th to the 9th century, with, for example, more people on a more restricted area.

Results and interpretations

The detailed study of the pottery has been focused on other aspects of the vessels than the origin of the pots. The study of the decoration was based on the classification system developed on the ceramics from Scharstorf (Meier 1990). This classification system was very useful in analysing the pottery from Groß Strömkendorf. One important result of the study is that a greater variety of decoration is present in the material from the central and southern areas. These areas are most likely the youngest within the trading site. The most common types of decoration are wavy lines and horizontal lines. Ceramics with such decorations were frequently found all over the site. Vessels with comb impressions were restricted to areas belonging to Phase II and III. Connections between particular decorations and rim types were also established. Rim type 22, which is the most common among the Sukow vessels, is only decorated in eight cases. For the remaining 112 decorated rim sherds, the most common rim type with a lip decoration is No. 25. This shape belongs primarily to the Feldberg vessels.

For the study of maximum grain size variation, measurements were taken directly on the sherds. As a result, it can be stated that the hand-made ceramics from the Baltic region are quite homogenous. This may be interpreted as representing fairly similar ceramic crafts. The grain size in wheel-thrown pottery differs more than 1 mm from the hand-made vessels. It is even difficult to identify any grains in some of the sherds from the Vorgebirge-Eifel region.

Regarding sherd thickness, the Sukow pottery differs from the rest of the Slavonic vessel types. Sukow is about 0.3–0.5 mm thicker in average. This could be the result of the turntable coming into use after the Sukow vessels; and, by using this tool, it was possible to make thinner vessels. But, it must also be stated that the differences are very small and could be an accidental result. The Scandinavian, Saxon and Frisian Coarse ware have about the same average sherd thickness as the Middle Slavonic types. Once again, the wheel-made pottery from the Vorgebirge-Eifel region differs significantly from the hand-made pots originating from the Baltic region. For instance, the vessel walls of Tating ceramics are about 4 mm thinner and the Badorf pots 3 mm thinner than the hand-made pots.

The study of the sizes of the vessels involved the height, rim diameter and base diameter. There are no clear differences between the various Slavonic vessel types. The main reason is probably that the statistical base is too small, as the younger material (Menkendorf and Fresendorf) is only represented by a few sherds. However, it has been noted that the Sukow and the Feldberg are about the same size. There are some discrepancies between the vessels from the cemetery and the ones from the settlement. The material from the cemetery is more homogenous than the settlement pots and large vessels were also only used at the latter. It seems likely that a grave vessel was usually of a more or less standardised size.

Almost 7,000 rims from Groß Strömkendorf were studied and classified according to the system established by Kempke (1981a). The study has shown that rim type No. 22 dominates on Sukow vessels and is also very common on Feldberg pottery. The main difference between the rim shapes of these two wares is that the lip is more rounded on Sukow pots than on Feldberg.

The shape of the bases was described according to the system developed by Meier (1994) on the Kosel material. At Groß Strömkendorf, most of the Slavonic vessels have flat bases with a sharp angular profile. Vessels with protruding foot show a connection to eastern Denmark (Scania). Measurements of the base thickness conducted on almost 60 vessels did not produce a result since the material was too scattered for any statistically reliable interpretation. Several of the vessels bear imprints on their bottoms from different materials or objects used as a support when the pots were built up, such as sand or organic material. Such support made it easier to rotate the vessel while forming it and also when removing it from the work surface. A few vessels show imprints from cloths or turntables. The cloth also may have had a support function. The cloth imprints were studied under the microscope in order to determine the weaving pattern. It appears that cloths in tabby and reps weave were used, both of which were common weaving patterns in Northern Europe during this time. The imprints from turntables are very unclear. The only certain example of a vessel with turntable imprints is a Menkendorf pot, dated to the second half of the 8th century or the beginning of the 9th century.

The grade of fragmentation of the pottery was calculated in order to study the intensity of using different areas at Groß Strömkendorf. The least fragmented sherds were found at the cemetery and in the north-eastern settlement area, while the most fragmented sherds derive from the other three settlement areas. This might indicate an increased intensity of activities at the trading site at the beginning of the 9th century.

Results of the scientific analyses of the pottery

Shape analyses

A total of 100 vessels with complete profiles from base to rim were discovered at Groß Strömkendorf *(App. 6)*. Only seven pots are not broken; the remaining pots were reconstructed from several sherds or have pieces missing. The analysed finds derive from every part of the settlement and from the cemetery. The dominating vessel types are Sukow with 64 vessels and Feldberg with 20 vessels. Only two pots are Menkendorf ware. The Scandinavian material consists of four vessels of which one is a hanging pot. The Saxon material comprises three vessels. The rest of the ceramics are Frisian, Merovingian Black ware, common Slavonic and one uncertain Ladoga vessel.

The pots were classified into 12 groups, named a–k and u. The latter group consists of five vessels of quite different shapes. No identical vessels were discovered among the excavated ceramics.

Shape group a

22 pots are classified as group a. The vessel profile is slightly S-shaped and the shoulder is situated at about 60 % of the height *(Fig. 40)*. The classified pots in this group mainly belong to Sukow and Feldberg, with one Scandinavian vessel. The latter has a somewhat shorter rim, but is nevertheless classified in group a. The height of Sukow vessels varies between 10 and 24 cm and of the Feldberg vessels between 12 and 22 cm. The Scandinavian vessel has a height of 14 cm. No correlation between shape and vessel type, maximum grain size, sherd thickness or rim type and decoration is apparent. The Sukow and the Feldberg pottery was discovered at the settlement and the cemetery. The Scandinavian and a Sukow pot were found in two different inhumation graves, which may have been connected to the Scandinavian burial custom *(see below)*. One hypothesis is that the deceased's relatives used a Slavonic vessel, resembling a Scandinavian one, in the grave, instead of making a new one.

Shape group b

Ten pots were classified in this group. The vessel profile is similar to group a, with two major differences *(Fig. 40)*: first, group b vessels have a slightly more rounded shoulder, and second, the shoulder is located slightly higher on the pot. The vessels in this group are of Sukow, Feldberg and Saxon types. The height of the six Sukow pots varies between 14 and 20 cm. The two Feldberg vessels are 14 cm in height and the two Saxon are 20 and 21 cm respectively. No correlation of the different parameters of the ceramics is observable. The Feldberg pots were used as urns in graves and the same applies for five of the Sukow vessels. The sixth pot was found in a pit. The Saxon material is from an urn grave and an uncertain find circumstance classified as a surface find.

Shape group c

The vessel profile of the four pots in group c is S-shaped with an everted rim *(Fig. 40)*. The major difference is the shape of the part between the base and the shoulder. In group c, this vessel part shows a smoother and more everted profile. In contrast, the ceramics in group a and b are more globular. Three of the vessels in this group belong to Sukow and the fourth is of Saxon origin. Statistically, the numbers are too low to allow any trustworthy interpretations, yet it is remarkable that all these pots were found as urns in graves. The height of the vessels varies between 9 and 15 cm, including the Saxon vessel with a height of 12 cm.

Shape group d

This group consists of 12 pots of Sukow and Feldberg ware. They are S-shaped with an everted rim *(Fig. 40)*. Apart from the fairly rounded shoulder, another characteristic of this group is the bottom shape: at the base, the profile is almost vertical and then curves outward. The Sukow and the Feldberg pottery in this group were discovered at the settlement and the cemetery. The cemetery pottery derives from seven urn graves. The variation of the vessel height is large. The smallest Sukow pot is 7 cm in height and the largest is 23 cm. The Feldberg pottery varies between 9 and 16 cm in height. Two of the Feldberg vessels are decorated with knobs. They are the only complete pots with knobs at the site, and were found at the settlement, in a pit and in a post hole. No other correlation between the shape groups, features and ceramics was observed.

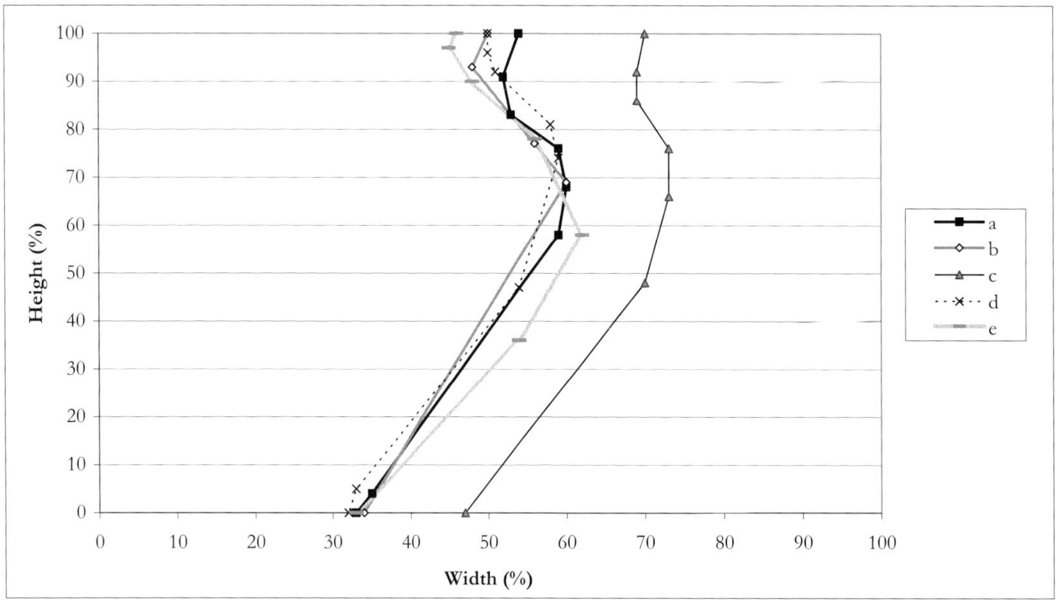

Fig. 40. *The shape analyses of group a-e at Groß Strömkendorf. One curve is representing a whole group, consisting of several vessels.*

Shape group e

15 Sukow pots, one Feldberg and one Slavonic pot were assigned to shape group e. The vessel profile is S-shaped with an extremely pronounced, rounded shoulder *(Fig. 40)*, looking almost like a bowl. The rim is very short and everted. Although these pots have several similarities with groups a and b, the differences in the shape are unmistakable. The Feldberg pot was found in some kind of structure of uncertain function. It has a height of 11 cm. The Sukow pots were discovered at the cemetery, as well as in various parts of the settlement. They belong to diverse types of settlement and grave features. The height of the Sukow pottery varies between 7 and 23 cm. The uncertain Slavonic pot has a height of 15 cm and it was found in a ditch at the settlement.

Shape group f

Group f is represented by six vessels, divided among Sukow, Feldberg, a possible Ladoga and a Frisian example. The profile of the vessels is S-shaped and quite similar to the pots in group e *(Fig. 41)*. The major difference is that the shoulder is even more pronounced and located more towards the middle of the vessel, at about 50 % of the total height. Vessels with rounded as well as distinct shoulders were attributed to this group. All pots were discovered in different kinds of graves. The uncertain Ladoga vessel was found in a mound, the Frisian vessel in a boat grave, the two Feldberg in urn graves and the two Sukow in an urn grave and in some kind of cremation grave. The height of the Ladoga and Frisian vessels is 10 and 25 cm respectively. The Feldberg vessels are 11 and 17 cm in height and the Sukow are 7 and 11 cm.

Shape group g

Seven pots were related to this group. All the vessels, except for one Feldberg pot, were classified as Sukow. Several Sukow pots were found in settlement features, while a few others as well as the Feldberg pot derive from urn graves. Typical of this group are the similarities with group f, but with one particular exception *(Fig. 41)*. The lower part, near the base, first turns about 45° outwards and at about 10 % of the height turns 75° outwards. The height of the Sukow pots seems to be more homogenous than that of the other types and varies between 10 and 15 cm. The Feldberg vessel is 9 cm in height.

Shape group h

Eight Slavonic vessels of Sukow and Feldberg type were classified as belonging to group h. Typical of the profile is the nearly bowl-like profile with a short rim turning outwards *(Fig. 41)*. The neck part above the shoulder is fairly long, measuring about 20 % of the total height. The Feldberg pots were found in urn graves. The Sukow pots derive from graves as well as from two pits in the north-eastern area and a post hole in the central area. The height of the Feldberg pots is 10 and 13 cm, and the Sukow pots are between 8 and 19 cm.

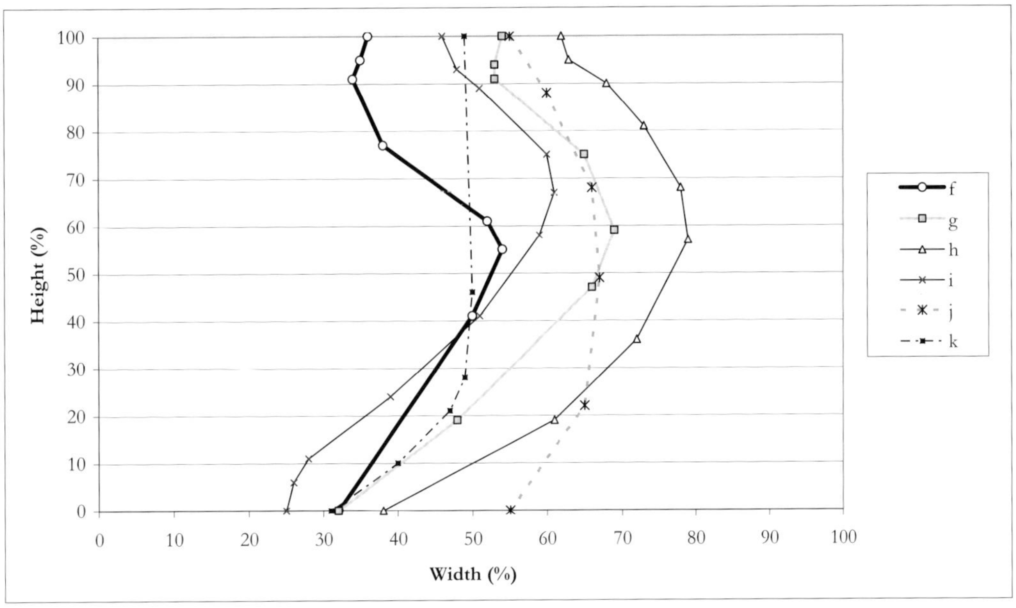

Fig. 41. The shape analyses of group f-k at Groß Strömkendorf. One curve is representing a whole group, consisting of several vessels.

Shape group i

Only two vessels were assigned to group i. Both are of Menkendorf type and were discovered in a mound and in some kind of uncertain feature. The vessel profile is straighter than on the other types, but it still has a distinct shoulder *(Fig. 41)*. The rim is more or less straight or turns slightly outwards. The height of the vessels is 11 and 14.5 cm respectively and the sherd thickness is 5 and 7 mm.

Shape group j

Two Scandinavian vessels and one of uncertain origin were classified as belonging to group j. The profile is similar to a bowl, with an inverted rim *(Fig. 41)*. The pots were discovered in three different pits, located in the central and north-eastern settlement areas. The Scandinavian vessels are 11 and 14 cm in height. The other vessel can be classified as a cup and is only 6 cm in height. It is clear that shape group j is of non-Slavonic origin.

Shape group k

Two vessels of Scandinavian and Frisian origin belong to group k. The profile is also bowl-like, but the shape of the rim is straight, instead of being inverted *(Fig. 41)*. The Scandinavian pot is 11 cm in height and looks, more or less, like a beaker. It was discovered in a well in the northern area, dendro-dated to 762. The Frisian vessel is about double the height of the Scandinavian pot and decorated with stamps. It was found in a pit in the northern area. This group is, similar to group j, of non-Slavonic origin.

Shape group u

Five Sukow vessels, which cannot be placed in other groups, were assigned to group u. Four of the vessels were settlement or surface finds. The fifth Sukow vessel was used as an urn. A vessel of Merovingian Black ware was added to this group *(Plate 24,1)*. The vessel was used as an urn, and there is no similar complete vessel at Groß Strömkendorf.

Results and interpretations

The shape analysis of the 100 vessels has resulted in some noteworthy conclusions. A fairly large amount of groups has been defined, but some of the groups are quite similar, with only very marginal differences. Groups a and b, which both contain a considerable amount of vessels, could perhaps be grouped together as one category. One of the most important results is that there are nearly no differences in shape between Sukow and Feldberg pottery. Both of these vessel types show the same distinct profiles. No special category, aimed for a specific type of ceramics, can be identified, and the relationship between the Sukow and Feldberg is evident. This is also confirmed by the similar vessel sizes.

Only two complete Menkendorf vessels were found, comprising a group of their own, group i. The Menkendorf pottery constitutes a new type of pottery with an unusual shape. The introduction of the Menkendorf pottery seems to be more obvious than the change from Sukow to Feldberg.

The relationship between the pottery from the cemetery and the settlement is clear. The same types and shapes of pottery were used as cemetery pots and settlement vessels. However, pots belonging to group c and f only were discovered in the cemetery area. These vessels are of quite different origins and were classified as Sukow, Feldberg, Saxon and Frisian ware. It can be argued that settlement vessels could have secondarily been used as cemetery pots. The natural science analyses are vital to prove such an interpretation. Another result concerns the ceramics in group j and k, which only contain imported pots from Scandinavia and Frisia. The differences with the Slavonic material are obvious. However, the analyses also revealed several similarities between the Slavonic, Saxon and other types of Scandinavian pottery.

A possible relationship between surface and rim decorations and the vessel profiles was evaluated. The only clear tendency is that a specific kind of knob on Feldberg only appears on vessels in group d. Meanwhile, no connection between the different shape groups and certain settlement areas and dendrochronological dated features is apparent.

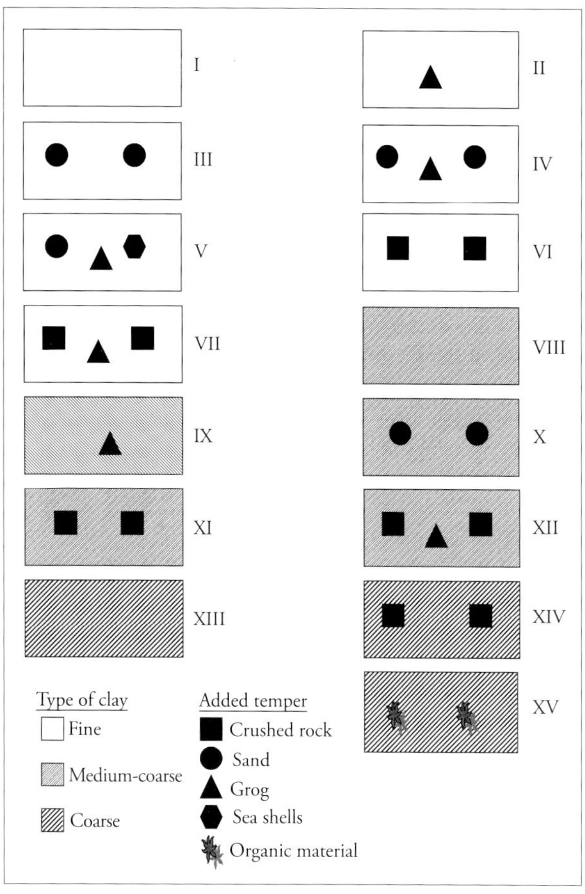

Several similarities and disparities were observed in the shape analyses. A central question is what affected the variation in the vessel shapes at Groß Strömkendorf more than 1200 years ago. The Sukow and Feldberg pottery were classified as belonging to several different groups. This could have had chronological reasons, but it seems more likely that they were produced contemporaneously. It is important to bear in mind that the ceramic material is extremely large and presumably represents a large amount of potters. No certain interpretation about the relationship between the individual vessels is notable. The only exceptions are the Menkendorf and the imported pots. It cannot be excluded that the same potter made several of the pots in one specific group.

Microscopic analyses

The 177 analysed sherds were divided into 15 ware groups, named I–XV *(Fig. 42; App. 7)*. Analyses also were conducted on a local clay sample. The coarseness of the clay and the type of temper were used as main parameters for dividing the ceramics. The coarse hand-made ceramics from Northern Europe and the wheel-thrown from the Vorgebirge-Eifel area have, to some extent, been grouped separately. The vessel building technique and the type of firing strongly point towards two different craft traditions. The ware groups illustrate the composition of the raw material used for making the pottery. The analyses partly confirmed the macroscopic recording of the vessel-building technique. The following presentation of the different ware groups is summarised because of the large amount of analysed sherds. Some specific and important details will be discussed after this presentation.

Pottery made from fine clays

Ware group I *(Fig. 43,1)*

The pottery in group I is represented by four wheel-thrown vessels, made from fine clay without any added temper. The largest grains measure from 0.5 mm to 1.3 mm. The sherds belong to two Tating vessels with tin-foil, one Merovingian Black ware and one Mayen sherd. One Tating sherd was discovered in a pit, the rest were surface finds.

Ware group II *(Fig. 43,2)*

The group only includes one sherd of a wheel-thrown vessel. The sherd consists of fine clay and the only added temper is grog in a small amount. The maximum grain size

Fig. 42. *Schematic description of the different ceramic ware groups at Groß Strömkendorf, identified through microscopical analyses of thin sections.*

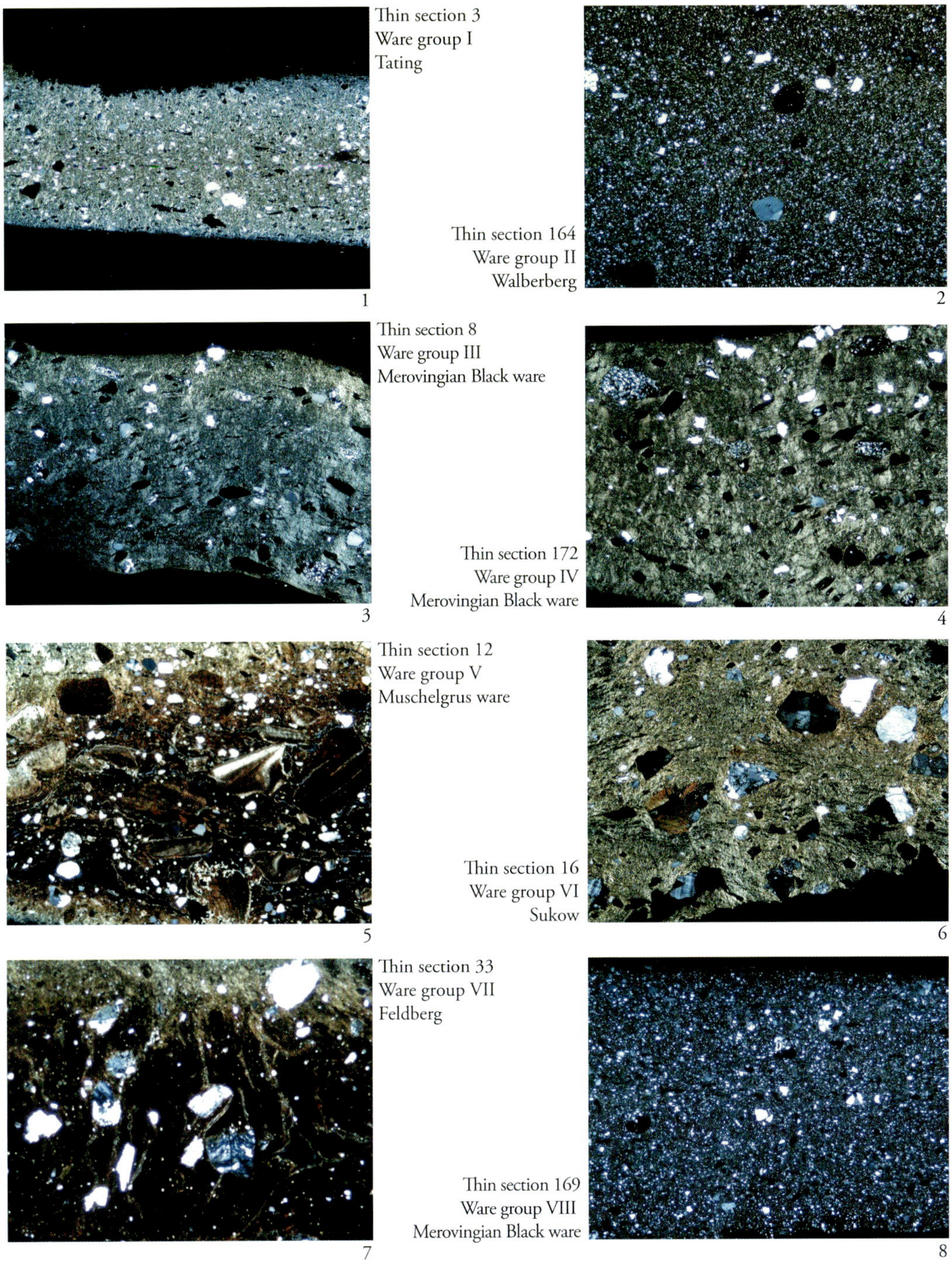

Fig. 43. *Micrographs of ceramic thin sections of pottery sherds from Groß Strömkendorf (photo: T. Brorsson).*

was measured to 1 mm. It is most likely a sherd of Walberberg type, discovered in a pit in the north-eastern area.

Ware group III *(Fig. 43,3)*

Group III comprises six hand-made and two wheel-thrown vessels, made from fine, sand-tempered clay. The hand-made vessels consist of four Frisian vessels discovered in a boat grave and in three different wells. Two sherds are of Sukow and Feldberg types, found in two pits. The amount of temper varies between 10 and 24 % and the maximum grain size is between 1 and 2.5 mm. The Frisian ceramics are slightly finer than the Slavonic ones.

The wheel-thrown pottery is quite similar and was classified as Merovingian Black ware. It derives from an urn grave and a pit. The amount of temper in both sherds is 5 % and the maximum grain size is 0.3 mm.

Ware group IV *(Fig. 43,4)*

A fine clay was used to make the four wheel-thrown vessels forming this group. Sand and grog in amounts of between 5 and 10 % were added as temper material. The maximum grain size was measured to between 1 and 1.5 mm. The sherds are classified as two of Merovingian Black ware, one Mayen and one Badorf sherd. They were found in three different pit houses and in a concentration of pits.

Ware group V *(Fig. 43,5)*

Eight hand-made vessels constitute this group. This type of ware is special and quite easy to identify with the naked eye, as the temper material consists of seashells. However, the fact that the clay is fine and that grog and sand additionally were used as temper is much more difficult to discover. The total amount of the temper varies between 10 and 20 % and the largest mineral grains are between 0.5 and 2.5 mm. This type of pottery mostly was classified as Muschelgrus ware, with one single sherd being probably of local origin. The analyses showed that this vessel also most likely belongs to the Muschelgrus ware. The pottery in group V is rather homogenous and may originate from the same region. At Groß Strömkendorf, the sherds were discovered in pits, in a pit house, in a well and as a surface find.

Ware group VI *(Fig. 43,6)*

The 69 analysed vessels belonging to this ware group were hand-made from fine clays. Various amounts of crushed rocks were used as temper, i.e., granite, gneiss and/or sand stone. Since the amount of material in this group is very large, the share of temper varies between 5 and 26 %. The maximum grain size is measured to between 1.5 and 5 mm. Most of the sherds in group VI were classified as Sukow, with Feldberg, Menkendorf, Fresendorf, uncertain Slavonic sherds and a lid appearing as well. The imported material consists only of Scandinavian sherds.

The ceramics were found all over the site in different kinds of graves and features at the settlement. Based on the large amount of sherds in this group it is most likely that using fine clays and tempering them with crushed rocks represents the common pottery technique at Groß Strömkendorf.

Ware group VII *(Fig. 43,7)*

The 17 hand-made vessels in this group were made from fine clays, tempered with crushed rocks and grog. The amount of temper varies between 8 and 16 %, and the maximum size of the grains between 1.5 and 4 mm. The vessels in this group were classified as Sukow, Feldberg, Scandinavian and Saxon. They mostly were found in pits, pit houses and in wells, but also as grave finds from urn graves.

Pottery made from medium-coarse clays

Ware group VIII *(Fig. 43,8)*

This group is represented by one hand-made sherd, seven wheel-thrown vessels and one crucible. They were all made from naturally tempered medium-coarse clays. Obviously, the potter did not add any other material to the natural clay. The hand-made sherd is, in fact, a handle *(Plate 26,9)* from an uncertain vessel type and was discovered in a pit. It is probably of non-Slavonic origin. The maximum grain size was measured to 1.25 mm. A similar handle was found in Ribe[5], in a layer dated to between 820 and 850. Another alike handle, which was placed on a bowl, comes from Dalem in Niedersachsen (Tiemeyer 1995, Taf. 23,1). The crucible is from a pit house in the central area. The maximum grain size is 1 mm. The practice of using naturally tempered clays for crucibles and moulds is very common (Stilborg 2003, 150 ff.). The piece from Groß Strömkendorf fits well into this picture.

The wheel-thrown sherds are divided into three Badorf sherds from pits and a dendro-dated well from 806, two Merovingian Black wares from a pit house and a pit, and two sherds of Tating without tinfoil. The latter sherds were discovered on the surface and in a pit.

Ware group IX *(Fig. 44,1)*

Only three sherds from wheel-thrown vessels are made from medium-coarse clays, tempered with small amounts of grog. The quantities are very small and no calculation of the amount was made. The maximum grain size measures to between 0.5 and 1.5 mm in diameter. Two sherds were classified as of Mayen origin, and the third as Merovingian Black ware. They were found in an urn grave, a well and in a ditch.

Ware group X *(Fig. 44,2)*

The one hand-made vessel and two wheel-thrown vessels forming this group were made from medium-coarse clays tempered with sand. The sherd representing the hand-made vessel is of Menkendorf type and was found in a pit in the central settlement area. The amount of sand in the ware is 12 %, and the maximum grain size is 2.5 mm.

The ware of the wheel-thrown vessels is much finer than that of the Menkendorf pot. The maximum grain size in both wheel-thrown vessels is 0.3 mm. The amount of temper is only 5 and 6 % respectively. Both sherds are classified as Badorf. One was found in a well and the other in a ditch.

---

5 The handle was studied during a visit in Ribe. It was found at the Post Office excavation.

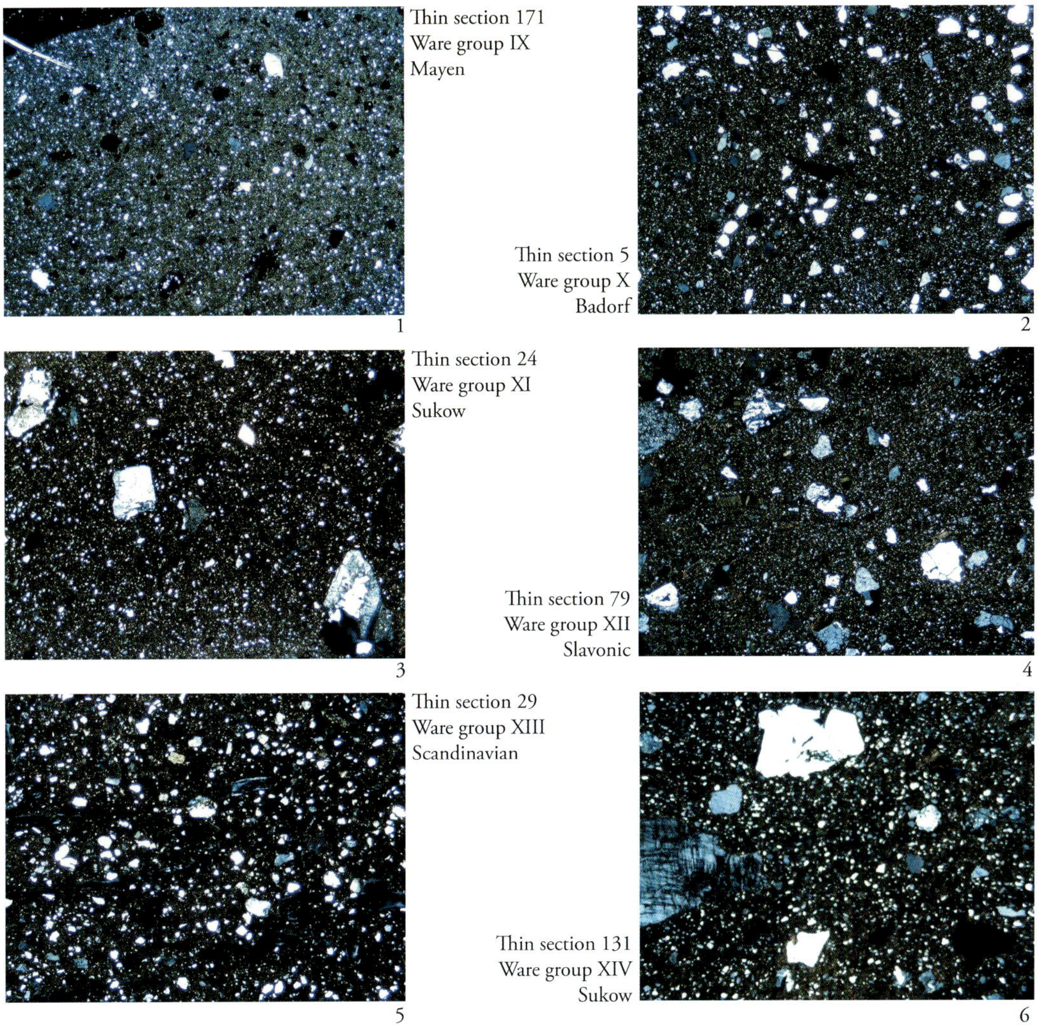

Fig. 44. *Micrographs of ceramic thin sections of pottery sherds from Groß Strömkendorf (photo: T. Brorsson).*

Ware group XI *(Fig. 44,3)*

Ten sherds from hand-made vessels were made from medium-coarse clays tempered with crushed rock. The identified rocks are granite and flint. The vessels are represented by three Sukow type sherds, four uncertain Slavonic type sherds, one possible Ladoga type, one Scandinavian as well as one possibly of Roman Iron Age. The amount of temper varies between 6 and 21 %, and the maximum grain size varies between 1.5 and 4.5 mm. The sherd dated to the Roman Iron Age was chosen as a reference to the older ceramics at the site *(Plate 26,5)*. The analyses showed that the vessel was made in the same way and with the same raw material as the Slavonic vessels, i.e., by using crushed rock as temper material. It fits well into the other ceramics in group XI. The maximum grain size is 3 mm and the amount of temper is 12 %. The pottery in this group was found in different types of graves and settlement features.

Ware group XII *(Fig. 44,4)*

Four of the analysed sherds from hand-made vessels were made from medium-coarse clays with crushed granite and grog used as temper. The temper varies between 9 and 22 %. The maximum size of the grains is between 2.5 and 3 mm. The vessels in the group are classified as two sherds of Sukow type, one Fresendorf and one uncertain Slavonic sherd. They were found in pit houses, a pit and as a surface find at the settlement.

Pottery made from coarse clays

Ware group XIII *(Fig. 44,5)*

33 sherds were made from coarse clays without any added temper. 28 of the sherds are hand-made and were classified as Sukow (eight sherds), Feldberg (five sherds), Scandinavian (twelve sherds), Frisian (two sherds) and a possible Curonian type (one sherd). The maximum

grain size in the ware was measured to between 0.5 and 3.5 mm. The finds are concentrated on the settlement and only three sherds were found at the cemetery. Two of these were located in urn graves and the third sherd, a Scandinavian, was used as an urn.

The remaining five ceramics are from wheel-thrown vessels, i.e., one Tating vessel without tinfoil, two Merovingian Black wares and two Badorf. The maximum grain size varies between 1 to 2.5 mm. The sherds were found in different settlement features.

Ware group XIV *(Fig. 44,6)*

Three of the analysed, hand-made vessel sherds were made from coarse clays tempered with crushed rock. The rock used is granite or sandstone. The amount of temper varies between 10 and 14 % and the maximum grain size is between 1.5 and 3 mm. The sherds were discovered as settlements finds in pits and in a pit house. There is one Sukow sherd, one Scandinavian sherd and an uncertain Slavonic type. The latter piece was sampled due to the rich amount of mica visible with the naked eye.

Ware group XV

The analysed sherd, which most likely belonged to a hand-made ember vessel, constitutes its own group. The vessel was made from a sorted coarse clay with organic material used as temper. The maximum mineral grain size in the ware was measured to 0.5 mm.

Local clay sample

A local clay sample was collected 75 m south of the southern settlement area. The clay is a sorted coarse clay with a maximum grain size of 1.5 mm. The mineral grains were mostly quartz, feldspar and mica.

Interpretation of the microscopic analyses

Pottery with temper of granite and/or sandstone

The dominating ceramic production technique seen in the pots at Groß Strömkendorf is one where the potter used different types of clay and tempered it with crushed rock. This technique is common in the Slavonic, Scandinavian and Saxon craft traditions. The mineralogical composition (both in clay and temper) in the ware of these vessels is also alike. It is mostly quartz, pieces of mica and feldspar, and grains of sandstone. The mineralogical structure in the different wares is quite similar and it is impossible to point out any specific production areas of the ceramics from the Baltic Sea region, since the same clays and rock minerals occur all over the southern Baltic region. Some odd grains, e.g., flint, were identified in a Feldberg and a Slavonic pot. Both these pots were used as urns. One is made from fine clay and the other from medium-coarse clay. Flint can also be found all over the Baltic region. In several of the analysed sherds, cracked or weathered grains were observed. It was rather easy to crush hearth stones or other stones that had been exposed to several heating and cooling events.

Pottery with sand temper

Sand temper mostly occurs in the pottery from the Vorgebirge-Eifel region. It is often associated with a medieval pottery craft, particularly with the use of the potter's wheel, kilns and glaze (Lindahl 2001, 295). It was part of a new ceramic manufacturing technique, based on other social activities than before. Sand-tempered vessels in the Baltic region also occur during prehistoric times, but are not as common as during the medieval time.

The mineral composition in these wares is similar to those tempered with crushed rocks. Sand temper was found in one single sherd of each, Menkendorf, Sukow and Feldberg. These sherds were located in three different pits in the central area. Three Frisian coarse sherds from three different wells were also tempered with sand. Two of these sherds are decorated with stamps. The fourth Frisian find is a complete vessel in boat grave No. 206. The Muschelgrus ware may also be connected to this group. The ceramics are tempered with sand, seashells and grog, but it seems likely that the Frisian pottery tradition was primarily based on sand temper.

Pottery with quartz-sand

The ware of two vessels classified as Frisian was tempered with quartz-sand. The grains are rounded. The mineralogical composition is mostly of quartz with a small amount of feldspar, which differs from the Slavonic ware. The proportion between quartz and feldspar has been used on medieval ceramics to point out different productions (Lindahl 1986, 117). In this regard, it is likely that the vessels from Groß Strömkendorf with quartz-sand temper have a foreign origin, Frisia being one of the most probable areas.

Pottery with grog

Grog was identified in several of the hand-made and wheel-thrown vessels at Groß Strömkendorf. The amount of grog is fairly low; only in the Muschelgrus ware it constitutes more than 5 % of the total temper material. The grog-tempered vessels were discovered at the cemetery and in all the different settlement areas, apart from the settlement in the north-eastern area. This can be explained by chronological circumstances, even though grog temper also appears during prehistoric times. The reason why the potter tempered the clay with a small amount of grog is not clear. These few solitary grains did not affect the strength, function or anything else of the vessel. It may instead have had other reasons beyond our knowledge. For instance, in present-day Sudan the potter includes one or two old sherds from her old, broken pot in order to give it a continued "life" in the new vessel (HULTHÉN 1985, 335).

Pottery with seashells – Muschelgrus ware

All the classified and analysed sherds of Muschelgrus ware are tempered with a large amount of seashells. The analyses were used to support the macroscopic recording. One of the sherds was recorded as being of probable local origin and two others as English Shelly ware. The analyses showed that all the sherds in group V consist of the same ware and that they are most likely of the same origin. There are in fact no indications of any English pottery among the analysed material.

Pottery with volcanic material – Mayen

Volcanic glass was identified in four sherds. Three of the sherds are classified as of Mayen origin and the fourth as Merovingian Black ware. The volcanic material in the clay supports the assumption that these vessels were, in fact, made from raw material from the Eifel region, where Mayen is situated.

Pottery with calciferous clays

Various amounts of calcium carbonate were identified in 11 of the 177 analysed sherds. Although the amount of calcium carbonate in the wares is fairly low, it may have affected the ceramics (*see page 84*). The sherds belong to different types of Slavonic pottery, and vessels of Scandinavian and Frisian origin.

Vessel building technique

The microscopic analyses have also shed light on the vessel building technique. The vessels classified as wheel-thrown were made from raw material consisting of clay and temper, which were worked together thoroughly. This ware gives a homogeneous impression, in contrast to the ware of the hand-made vessels. The grains of temper in the latter ware are occasionally concentrated on a few parts of the matrix, resulting in a more or less heterogeneous pattern. The identified vessel building technique in all the hand-made vessels is coiling. This is visualised in thin section No. 33 *(Fig. 43,7)*. Parallel pores running diagonally from the outer to the inner surface were identified under the microscope. The pores are the contact zones between different coils of clay. The coils were most likely joined by the so-called N-technique[6]. It means that the clay of the coils was in turns pulled upwards on one side and downwards on the other side. The illustrated thin section belongs to a Feldberg vessel with a burnished surface and wavy-line decoration. The sherd was found in a pit in the central settlement area, and is most likely dated to the end of the 8th century.

The origin of the pottery

In the study of the origin of the ceramics, it is conducive to combine the microscopic analyses with a survey of the soil. The geological structure of the soil in this part of the northwestern region of Mecklenburg displays different fractions of sand and moraines (GEOLOGISCHE KARTE VON MECKLENBURG-VORPOMMERN). The soil of Groß Strömkendorf and its surroundings consists mainly of sand, but there are also several areas with peat and peat bogs. According to the geological map, no deposits of clay were identified in the area. However, clay was found about 75 m south of the southern settlement area, near the shoreline. Analyses of a clay sample have shown that it is a coarse clay, rich in quartz, feldspar and mica. The clay is of the same type as that belonging to ware groups XIII and XIV. The majority of the Slavonic vessels were made by finer clay types, which must have been collected at other places. The vessels in groups XIII and XIV may have been made from the same clay as the local sample, but they could also have been made from raw material from, for example, the other side of the Baltic Sea. It can be stated that the mineral composition of the local clay sample is the same as the majority of the analysed sherds from Groß Strömkendorf.

Most of the different Slavonic vessels at Groß Strömkendorf were, as mentioned above, made of fine clays and tempered with crushed rock. Among the few vessels, which were instead made of medium-coarse or coarse clays, are Sukow, Feldberg and Menkendorf types. The coarseness of the clay may be an indicator of the origin of the vessel, but it can also point out different productions within the same society or perhaps different vessel functions. The mineralogical composition of the ware is an important contribution to this discussion. All analysed Slavonic vessels at Groß Strömkendorf have more or less the same composition providing a fairly strong evidence for a local production. However, as mentioned above, the mineralogical composition in the entire southern Baltic region is more or less the same. The presence of diatoms could also shed light on the discussion of the origin of the pottery. Unfortunately, no diatoms were identified in any of the analysed sherds. To conclude, the analyses of the thin sections cannot ascertain a specific production area for the Slavonic pottery.

6 ANDERSEN 1975, 56 f. The method of joining the coils was established on the basis of the Ertebølle ceramics from Ringkloster, Jutland, Denmark.

Only two sherds of Saxon origin were studied and they were made in the same way. The vessels may be of the same origin. Among the Frisian Coarse ware, two different groups appear. These groups may reflect different functions or perhaps different origins. The third Frisian group is the Muschelgrus ware with a most certain common origin.

The question about the origin of the wheel-thrown vessels is also uncertain. Altogether, six Badorf sherds were analysed, comprising three different ware groups. The same pattern is noticed among the Mayen and Merovingian Black ware. It is obvious that several different productions existed within every pottery group. Volcanic glass was identified in three of the four sherds of Mayen type. These three sherds belong to three different ware groups. This result is the same as that arrived at by Stilke et al. (2001) when using neutron activation analyses. Even here, several groups of Mayen ceramics were identified.

One question that has been much debated is the origin of the Tating ware. The five sherds from Groß Strömkendorf were divided into three different ware groups. The two sherds of black burnished ware with tinfoil belong to one group and the two sherds of greyish ware to another. It is obvious that two different productions are represented, which may support the result by Hodges (1981, 65) *(see page 38)*. The fifth sherd is also black burnished but without tinfoil. This sherd does not belong to any of the others and constitutes a separate category. A connection with the Mayen production was not supported by the analyses of the Tating ware from Groß Strömkendorf. The main reason for this is the lack of volcanic material. The analysis of the Tating ware did not give any new information about the function of the jugs, as either liturgical or a more common object used in the household.

Grain size distribution analyses

A method for analysing ceramic material is measuring the grain size distribution. The main differences between this analysis and, for instance, microscopy of the ware is the investigation of the small fractions, especially the silt fraction. The grain size analysis does not involve the mineralogical composition of the ware. It is only focused on the size and distribution of the grains. It is a method for quantifying the coarseness of the clay and, when applicable, the temper.

The amount of grains and their sizes were calculated in two separate fractions. The first calculation was performed on grains smaller than 0.1 mm but larger than 0.01 mm in diameter. The second calculation was made on grains larger and equal to 0.1 mm. In all thin sections four calculations were made for each fraction. The area covered with grains, both in mm^2 and percentage, as well as the number of grains in every fraction was recorded and the mean value was calculated. By working with this method it is possible to gain an idea about the size of the grains in a specific fraction. A higher value indicates to some extent small grains whereas a lower value indicates larger grains.

Number of grains – grain size

A correlation between the mean value of the number of grains and the mean grain size was carried out on a selection of thin sections. The analysis was made on fractions between 0.01 and 0.1 mm, as well as on fractions larger than 0.1 mm. It can firstly be stated that the material seems to be fairly homogenous. Only a few sherds differ from the rest of the material.

In the smaller fraction, eight sherds in particular are different. These are the sample Nos. 66 *(Plate 21,5)*, 85, 106 *(Plate 13,5)*, 163, 69, 104 *(Plate 13,4)*, 115 *(Plate 5,17)* and 164. The sherds were classified as Sukow, Feldberg, Menkendorf, Jutlandic, Walberberg and Mayen. Several of the Slavonic sherds have quite small grains in this fraction. This could be the result of fine clays being used for the Slavonic pottery.

The pottery with the fraction of grains larger than 0.1 mm is also quite homogenous. Only five sherds differ considerably from the rest: sherds Nos. 1 *(Plate 23,1)*, 57 *(Plate 21,4)*, 29, 4 and 6 *(Plate 25,4)*. These sherds were classified as Frisian Coarse ware, Scandinavian, Merovingian Black ware and Badorf. It is notable that no Slavonic sherd is among this material. The grain size distribution of the larger fraction can be used as an indicator of foreign ceramics. Another result is the connection between samples Nos. 162 *(Plate 25,2)* and 163, classified as Mayen. These wares are very similar. The same applies to samples Nos. 164 and 165, which were classified as Walberberg and Mayen.

Number of grains – area in different fractions

The average number of grains in the fractions between 0.01 and 0.1 mm and the fractions larger or equal to 0.1 mm were also compared. The correlation shows that the material is distributed over a wide range. The majority of the sherds belong to a cluster and are almost inseparable from each other. Most of these sherds belong to different Slavonic types and are comprised especially of wares of fine or medium-coarse clays. Seven sherds differ significantly from the rest of the ceramics. A group consisting of five sherds classified as Sukow (No. 103), Scandinavian (No. 29), Walberberg (No. 164), Badorf (No. 6, *Plate 25,4*) and Merovingian Black ware (No. 4) can be classified as rather coarse ware. The microscopy of the wares showed that the Badorf and Merovingian Black ware were made from medium-coarse clay and the Sukow and Scandinavian from coarse clays. The reason why the Walberberg sherd, a fine ware, is located among pottery of coarse ware can be explained by the grain size distribution. The smallest grains among the analysed material were identified in the Walberberg sherd. The amount of grains compensated the grain size.

Another sherd that differs from the rest of the material is a sherd of Sukow type, No. 115 *(Plate 5,17)*. The ware consists of large amounts of grains in the smaller fraction, and only a minor amount of grains in the larger fraction, i.e., larger than 0.1 mm. The microscopy of the ware showed that this vessel was made of medium-coarse clay, tempered with 12 % crushed rock. Sherd No. 2, a Tating vessel, only contains a few grains both in the smaller and larger fraction. The microscopy of the ware revealed that the pot was made of fine, sorted clay without any added temper. The Feldberg sample No. 63 *(Plate 10,1)* also has a small amount of grains in the smaller fraction but an average distribution in the fraction larger than 0.1 mm. The temper material, which consists of crushed rock and grog, most likely affected the result of the grain size analyses. Sherd No. 1, which is a Frisian Coarse ware, constitutes a separate group in the plot. It is clear that the vessel was made from another raw material than the Slavonic ware.

In many ways, the grain size distribution supports the microscopy of the ware. A problem is that only a few sherds were studied by both grain size distribution and microscopy of the ware. But the result can be used as an example for future research.

Only one sherd of Badorf, one of Walberberg and one of Tating were studied. The grain size distribution of the number of grains placed these vessels into separate areas in the diagram. There are no similarities between the wares of the Tating, Badorf and Walberberg sherds. It is very clear that different raw material sources were used when making these pots. The Merovingian Black ware as well as the Mayen sherds are scattered over a wide section of the plot. It seems likely that the analysed sherds represent several different productions from the same region.

Two Sukow vessels differ from the rest of the Slavonic material. They are sherd No. 115, which is made of a medium-coarse clay, and sherd No. 103, made of a coarse clay. The latter sherd was found in a well that was dendro-dated to 809. These two sherds were made from other clays than the majority of the ceramics, and thus could have originated from other places than the rest.

Thermal analyses

Thermal analyses were carried out on a local clay sample and on a total of 25 pottery sherds from the trading site and the cemetery *(Tab. 23)*. The samples represent vessels

TCT Group	Find number	Type of feature	Area	Vessel type	Dating	Thin section	Firing temperature °C
I	98/125/632,235	Well	Southern	Badorf	806	160	1125
I	97/88/10115,143	Ditch		Badorf		6	1125
II	97/87/215,307	Pit house	Central	Mayen		162	1100
II	90/314,426 ,427			Merovingian		173	1100
II	98/125/242,445	Pit house	Central	Merovingian		4	1100
III	98/126/800,1	Urn grave		Merovingian		7	1225
III	97/87/256,130	Pit - concentration		Merovingian		172	1175
IV	96/518,6	Surface find		Mayen		163	1200
V	97/87/256,118	Pit - concentration		Tating		167	1175
V	97/87/276,120	Pit		Tating		3	1175
VI	98/125/723,19	Pit		Muschelgrus		14	750
VI	97/87/370,14	Pit house	Southern	Muschelgrus		13	750
VII	95/541/1,70	Pit house	Central	Sukow		18	850
VII	95/541/28,41a	Pit house?		Feldberg		21	950
VII	96/302/59,61	Pit		Scandinavian		29	950
VII	98/126/753a,1	Urn grave		Sukow		60	850
VII	98/126/782d,1	Urn grave		Feldberg		64	850
VII	98/126/1034,1	Pit		Sukow		154	850
VIII	97/87/313,324	Well	Southern	Scandinavian	809	108	850
IX	96/302/62,202	Pit		Feldberg		33	950
IX	95/541/1,105	Pit house	Central	Sukow		20	850
IX	98/126/10370,89	Well	Northern	Scandinavian	761	135	850
X	98/125/656,205	Well	Southern	Sukow		62	950
X	97/87/313,243	Well	Southern	Feldberg	809	106	950
X	97/87/313,324	Well	Southern	Sukow	809	107	850

Tab. 23. *The result of the thermal analyses. The ceramics have been divided into ten different groups according to the firing temperature and the character of the ware.*

from the Slavonic region, Scandinavia, the Frisian coastal area and the Vorgebirge-Eifel region. The Slavonic sherds are of Sukow and Feldberg type.

The analysed material can be divided into two groups. One group is the hand-made pottery from Northern Europe, which also includes the Frisian coast. The other group comprises wheel-thrown pottery from the Vorgebirge-Eifel region. The hand-made pottery is fired at a temperature below 1000°C and the wheel-thrown vessels at temperatures exceeding 1100°C.

A definition of ceramic ware, in order to distinguish it from raw clay, is that it will keep its shape even when it is soaked in water. This point is reached when the firing temperature exceeds about 450°C. At this temperature, the crystal water of the clay minerals begins to evaporate. Thus the original crystal structure is destroyed.

Clay sample

Thermal analysis was carried out on a sample of raw clay, collected 75 m south of the southern settlement area. The aim of the study was to determine the sintering and melting point of the clay. Generally, it started to sinter between 1100° and 1150°C and the melting point was reached at 1300°C.

Hand-made pottery (temperatures below 1000°C)

Slavonic pottery *(Fig. 45,1)*

A Thermal Colour Test (TCT) was conducted on six Sukow sherds. One sherd started to change colour already at 300°C, and the rest changed from 400°C, which was below the original firing temperature. Five of the six examined Sukow sherds started then to change colour between 800°C and 900°C, which is most likely the original firing temperature. Another sherd, No. 62 *(Plate 4,9)* was almost certainly fired to between 900°C and 1000°C. One of the Sukow sherds (No. 18, *Plate 6,17*) started to sinter around 1150°C, and the rest between 1150°C and 1200°C. All the analysed Sukow sherds melted at the same temperature, around 1300°C.

Four sherds of Feldberg type were analysed. One sherd, No. 64 *(Plate 9,3)* started to change colour at 300°C, another, No. 21 *(Plate 12,2)*, at 400°C, and Nos. 33 and 106 at 500°C. Nos. 64 and 33 are the vessels, which, in all probability, were fired at the lowest temperatures, between 800°C and 900°C. The others were fired at between 900°C and 1000°C. The lowest sintering point, around 1150°C, was yielded for ware No. 33. The sintering point of ware No. 21 was reached at 1200°C and for the others at 1250°C. The melting point of ware Nos. 21 and 33 is 1300°C. The other Feldberg sherds started to melt at 1350°C.

Some sherds of Sukow and Feldberg were fired to between 800°C and 900°C, while others of the same type were exposed to an additional 100°C. The sintering and melting point of the Slavonic pottery also varies. It seems likely that different clay and temper materials are represented in the analysed objects. The Sukow sherds and two of the Feldberg sherds can be connected to the local clay sample from Groß Strömkendorf. Some minor differences can be observed but it seems possible that several of the Slavonic vessels were made from the same type of clay as the local clay sample. An important result is the fact that there are almost no differences in the original firing temperature between the Sukow and Feldberg pottery. It seems probable that the vessel types were fired under the same conditions.

Scandinavian pottery

Thermal analysis was performed on three sherds of Scandinavian type vessels. The sherds started to change colour at 400°C. Sample No. 108 as well as sample No. 135 changed colour between 700°C and 800°C. Sample No. 29 was fired to an additional 50–100°C. These temperatures represent, in all likelihood, the original firing temperatures. Sample No. 135 started to sinter at 1150°C, and samples No. 29 and 108 at 1200°C. The sherds were completely melted at 1300°C. The thermal analyses indicate that the sherds consist of nearly identical wares. Microscope analysis showed that sample No. 135 differs from the other two Scandinavian sherds. The sintering point of these wares supports this result. It may point towards separate origins of the Scandinavian sherds, but a common craft tradition with approximately the same firing temperatures.

Shell-tempered pottery

Two sherds of shell-tempered pottery (No. 13 *[Plate 23,7]* and 14) underwent thermal analyses. The sherds behaved exactly the same. At about 500°C, minor changes in the colour were observed. A major change appeared between 700°C and 800°C. This temperature is most likely the maximum temperature that the wares previously were exposed to. At 1200°C, both samples started to sinter, and the melting point was reached at about 1300°C. The sherds are almost certainly from the same pottery production or workshop and made from the same clay type and temper material.

Wheel-thrown pottery (temperatures exceeding 1100°C)

Badorf pottery

Thermal analysis was performed on two sherds, classified as Badorf (samples Nos. 6 and 160). They behaved nearly the same at the low temperatures. A minor change in colour was observed around 600°C but it is not likely

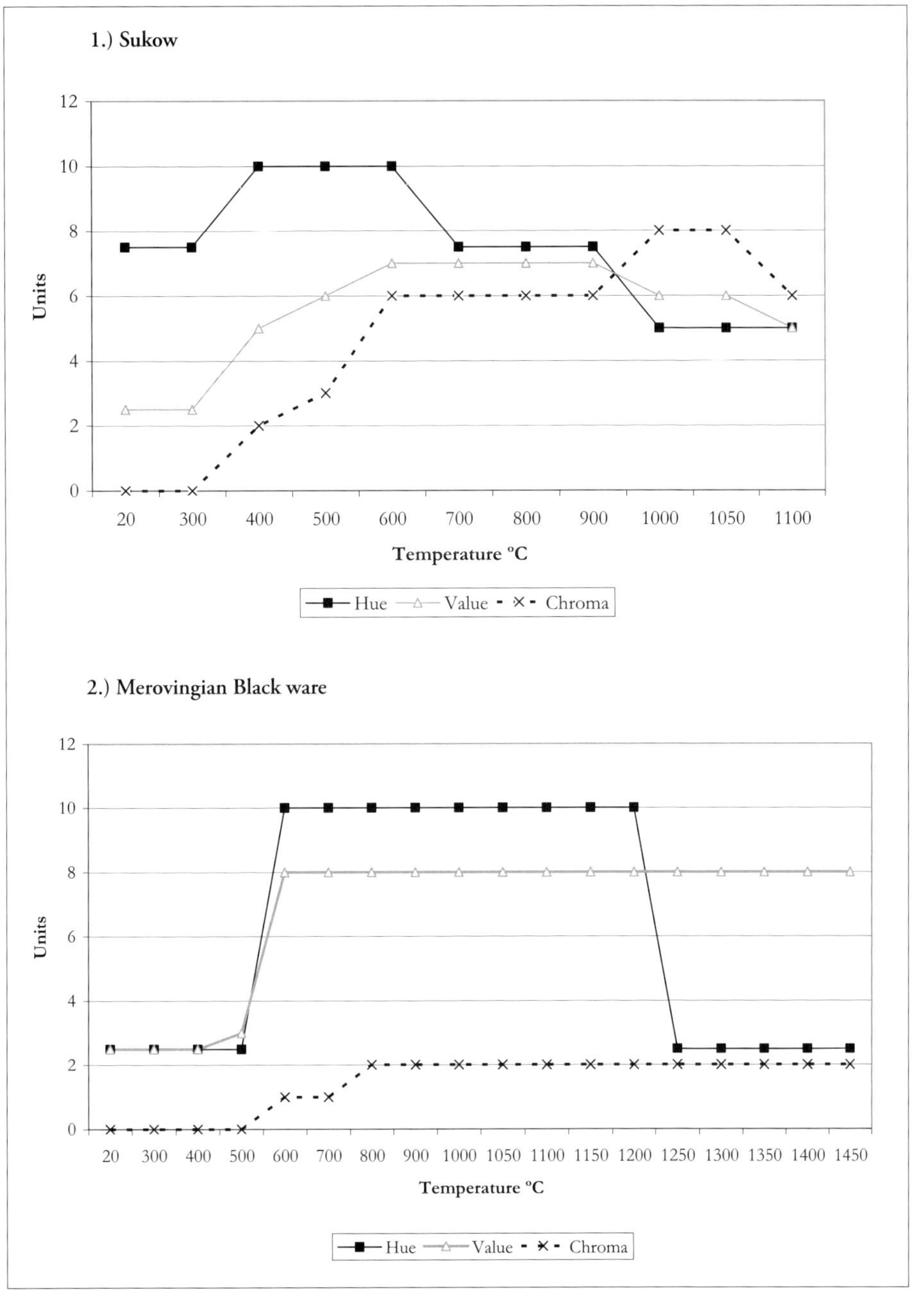

Fig. 45. Plots of samples of the Thermal Colour Test groups identified at Groß Strömkendorf. 1.) Sukow; 2.) Merovingian Black ware.

the original firing temperature. This point is most likely represented by the colour changes between 1100°C and 1150°C. The sintering point was extremely high, at 1400°C. The melting point was reached at 1500°C.

Merovingian Black ware *(Fig. 45,2)*

The Merovingian Black ware sample consisted of four analysed sherds. The pottery did not behave exactly the same at the lower temperatures, below 1000°C. Minor changes were noticed but these were without any doubt not caused by the original firing temperature. The first major change in the colour appeared between 1050°C and 1100°C in samples Nos. 4 and 173 *(Plate 24,2)*. This was most likely their original firing temperature. Samples No. 7 *(Plate 24,1)* and No. 172 *(Plate 24,3)* seem to have been fired to between 1150°C to 1200°C and 1200° to 1250°C respectively. The sintering point was reached at different temperatures, from 1350°C in sample No. 4 to just below 1500°C in samples Nos. 7 and 172. The melting point was reached at an additional 50°C.

The four analysed sherds comprise three different thermal groups. It seems possible that sample No. 4 and No. 173 were fired at the same occasion but the sintering and melting points indicate two different raw materials. The other two analysed sherds show partly the opposite picture. They were not fired at the same temperatures, but the raw material could be the same. According to the archaeological classification, it is possible, that the vessels are from the same region and pottery tradition.

Mayen pottery

The TCT analyses indicate that the two sherds interpreted as originating from Mayen, No. 162 and No. 163, were fired at different temperatures. Below 1000°C, only minor changes in colour were observed. Above 1000°C the colour changes were more significant. However, they were not the same for the two samples. It is estimated that sample No. 162 was originally fired to around 1100°C and sample No. 163 to approximately 1200°C. The sintering point in both samples was reached at 1350°C and they were completely melted at 1500°C. The Mayen sherds seem to have been made from the same raw material but were fired at two different temperatures.

Tating pottery

Two sherds of Tating pottery, a black burnished sherd and a greyish sherd, were analysed. At the low temperatures, the sherds behaved nearly the same. A major change in colour occurred between 1100°C and 1150°C, which is most likely the original firing temperature. The sintering point was reached at 1300°C and the melting point at 1400°C. The sherds are very similar and most probably from the same pottery tradition.

Results and interpretations of the TCT analyses

The thermal analyses clearly show the differences between hand-made pottery types from the Baltic region, Saxony and Frisia. The hand-made pottery seems to have been fired in pits or in bonfires at temperatures between 750°C and 900°C. However, the Slavonic pottery was fired to around 900°C, which is a very high temperature for a fire in a pit or a bonfire. It cannot be ruled out that this type of pottery was instead fired in some kind of kiln. There are no differences between the Slavonic pottery types and it seems likely that the same technique for firing ceramics was used during the whole time the settlement was in use. There also are no differences between vessels from the trading site and vessels used as grave goods. For the Scandinavian material two separate origins are indicated, which is also supported by the microscopic analyses. The Scandinavian sherds seem to have been fired at lower temperatures than the Slavonic pottery.

The two sherds with shell temper constitute their own group. They are most likely from the same pottery tradition and perhaps made at the same place and even fired on the same occasion.

A problem when analysing ceramics, fired in bonfires or in pits, is the broad variation of maximum temperatures, even during the same firing. Thermometric measurements were made on material collected from ethnoarchaeological studies. The measurements have shown that the variation inside different kinds of firing structures could be significant, normally around 300°C (Gosselain / Livingstone Smith 1995, 154 Fig. 5). During firing, the vessels are placed in separate parts within the fire and can thus be fired to various temperatures, depending on the size and amount of firewood and the wind conditions. It is therefore important to use the results of thermal analyses with care.

The vessels from the Vorgebirge-Eifel region can be classified as almost as hard as Stone ware pottery. They were originally fired to between 1050°C and 1200°C. The clay was carefully collected for this purpose, and some of the clays did not melt even at 1500°C, which is an extremely high temperature. The thermal analyses clearly showed why the Vorgebirge-Eifel region became an important ceramics centre, producing pottery for hundreds of years and distributing it to several European countries. The analyses also revealed that several productions are represented among the material from Groß Strömkendorf.

Concerning the Tating ware, it is notable that the two analysed sherds with burnished and greyish surfaces respectively were possibly made from raw material from the same area. This result is important in the discussion about the origin of the Tating ware. The Merovingian Black ware is divided into several pottery productions. Sherds No. 7 and No. 172 most likely were made from the same raw material. The microscope analyses showed that both vessels were made from fine clays but with different temper. In one of the pots, grog as well as volcanic material was identified. Both clays were tempered with sand. The thermal analyses as well as the grain size distribution analyses indicate that there is a connection between the Merovingian Black ware and the Mayen ware. It is likely that the productions of these two ware types were located in the same region.

Ceramic craft at Groß Strömkendorf

The ceramic production during the 8th and early 9th century at Groß Strömkendorf was based on a craft tradition that was several thousand years old. The ways of processing the clay and building the vessels were more or less stable from the Stone Age up to the 8th century. One of the first major changes in the method of making pottery occurred at this time. It was the same period during which the settlement at Groß Strömkendorf existed. A new tool for making pottery was introduced – the turntable. The pottery during the 8th and early part of 9th century was most likely made on a household level. The question is if the later Middle Slavonic ceramics were made in special workshops or not.

Several ethnographic studies have shown that most of the pottery produced in the households was made by women (Stilborg 2002, 36 f.). This pottery was used in the family, perhaps a few vessels were sold to other families. They could have been traded for food or other objects. Vessel production in workshops was normally associated with the use of kilns and special tools, e.g., the potter's wheel (ibid. 36 f.). All the Slavonic pottery at Groß Strömkendorf should, according to this background, have been made in the households.

In order to study the different aspects of the ceramic production at Groß Strömkendorf, all different steps of making vessels will be described, from the selection of raw material to the finished pot.

Raw material

The choice of raw material for the production of pottery was affected by several different factors. For instance, the quality of the clay is based on the mineral composition, plasticity, particle size, the amount of soluble salts and non-plastics present (Arnold 1985, 21). Most of the Slavonic pottery at Groß Strömkendorf was made from glacial, sorted fine clays, containing small amounts of silt and almost no sand. No diatoms were observed in the clay. The amount of mica in the clays varies between poor and very rich, which indicates that clay was most likely collected from several different sources. Depending on the depth of a clay pit, the amount of mica and other inclusions may vary.

One question is how far from Groß Strömkendorf the clay was collected. Several different methods for analysing ceramics have been used in archaeological research but there is still no way to determine the exact origin of the clay and the ceramics from the Baltic region. The main reason lies, as mentioned above, in the soil deposition caused by the ice from the latest glaciation. When the ice retreated to the north, it left behind soils and stones of mixed origins, resulting in the fact that the geology in northern Germany and Poland is very similar to that of southern Scandinavia. In the absence of written sources from the Early Medieval time, ethnographic studies have been one important source for the study of pottery craft. However, ethnographic parallels are connected with a number of source-critical aspects. They only can offer suggestions about the prehistoric society and the ceramic craft. Studies from 20th century-Peru (Arnold 1993, 67) and Zimbabwe (Lindahl / Matenga 1995, 27) have shown that people went by foot not more than 2 or 3 kilometres to obtain clay and that a potter normally used the same clay source until it was depleted. At Quinua in Peru the ceramists bought or obtained the clay by other methods from landowners who owned areas with clay deposits. The ethnographic parallels from Peru also have described that the potter sometimes chose special gold-like clays (Arnold 1993, 73). These clays were rich in mica, and the potter knew that they were very good as a raw material for vessels used for cooking and brewing purposes. Other clays without the gold-like particles were made into pottery for carrying water, storing food, for rituals and for tourists. At Groß Strömkendorf, a special group of pottery has been noted to be rich in mica. The analyses of three sherds of a Sukow (thin section 158), a Feldberg (thin section 159) and a pot of uncertain origin (thin section 156) indicated that these clays were, in fact, extremely rich in mica. Two of the sherds were found as surface finds at the southern and central settlement and the third was a find from a pit house in the north-eastern area. The Sukow and Feldberg vessels belong to ware group XIII. They are made of unsorted coarse clays without any added temper. It is a choice of raw material, which was mainly associated with the Scandinavian pottery tradition at Groß Strömkendorf. The sherd of uncertain origin belongs to ware group XIV (coarse clay with crushed rocks used as temper).

Another ethnographic parallel comes from Zimbabwe and concerns the selection of clay (LINDAHL / MATENGA 1995, 29). One potter used two different clays. A fine clay was applied for making small vessels like cooking and drinking pots. When the potter tried to make large containers of that clay they collapsed during the process. Instead, a coarse clay was well suited for the manufacture of both small and large vessels. The potter replied to the question why she did not use that specific clay for all the pots: "Don't you understand that such a coarse clay is not adequate for cooking pots, the large grains come off when you stir the food and you don't want gravel in what you are eating, do you?".

The clay used for the ceramics at Groß Strömkendorf was mainly non-calciferous. Nevertheless, in a few sherds calcium carbonate was identified. The amount is fairly low and most likely did not affect the ware. The person who made the vessel tried with all certainty to avoid calcium carbonate in the clay. This substance can cause problems during and after firing. The calcium carbonate expands after firing, which can break the vessel. This phenomenon can be avoided if the firing is reduced or organic material is mixed into the clay. The organic material in the ware generates a carbonisation, which has the same effect as a reduced firing. Organic material in the thin sections is observed in carbonised remains in the holes in the ware. It is often difficult to identify the kind of organic material. Most common are different types of plant material such as grass, roots, seeds, chaff and leaves. The latter was identified in an uncertain Curonian sherd, thin section No. 146 *(Plate 26,4)*. Small amounts of organic material were identified in a total of 25 sherds of quite different origin and from different areas. None of these vessels was found at the cemetery. The clays used for the Slavonic pots at Groß Strömkendorf were mainly tempered with crushed granite and/or sandstone in different proportions. The grain size varies between 1.5 and 5 mm. It is most likely that stones that had already been fired were used, since they are much easier to crush than a raw stone. Such pre-fired stones can even be crushed by hand. Several of the imported vessels at Groß Strömkendorf were made of sand-tempered clays. Among the analysed material, it appears that only one Sukow, one Feldberg and one Menkendorf type sherd were tempered in this way. These sherds were discovered in three different pits in the central area. No sand-tempered pottery was found associated with the oldest settlement phase. It is possible that the foreign ceramics may have affected the local Slavonic craft over time.

The stones used as temper were in all likelihood collected in the surrounding area by the potter. However, it is possible that even special temper mines were in use, as, for example, ethnographic studies from Peru have shown (ARNOLD 1993, 64 ff.). There, a few persons worked at such mines and sold the temper material to several small villages within a radius of a couple of kilometres from the mines.

Vessel building technique

The pots originating from the Baltic Sea and the North Sea regions were manufactured in a coiling or modelling technique, whereas the pots from the Vorgebirge-Eifel area were made on the potter's wheel. The use of the potter's wheel made it possible to make more pots within a shorter time. This tool was one part of the technological development of the pottery craft. The potter's wheel also facilitated some steps in the decoration and finishing process. There is no evidence of any local production of wheel-made pottery among the Groß Strömkendorf material.

The Sukow pots were made by hand and most of the vessels were built up by coiling *(Fig. 46)*. However, some sherds show characteristic finger marks, indicating that modelling was used as well. Altogether, 17 Sukow pots were probably shaped on turntables. Characteristic horizontal lines on the inner and outer surfaces of the upper part of these pots were identified. The differences between pots shaped on turntables and pots made on the potter's wheel are mainly seen in two important aspects of the pottery. The first concerns the temper: large and angular grains in the clay used for wheel-throwing could hurt the potter's hands. The second aspect concerns the presence of the horizontal lines on the inside of the vessels. Horizontal lines appear on wheel-thrown pots on the whole inside, from the base to the rim. On pots shaped on a turntable, lines occur only on the upper part. The constructions of the potter's wheel and the turntable are quite similar, but differ in one important aspect (*Fig. 47*). The potter's wheel is rotated by foot or by stick, whereas the turntable is rotated by hand. Thus, the rotation speed is much higher on a potter's wheel. Practical experiments have shown that it must be higher than 0.7 m/sec. Otherwise the vessel wall will collapse and make the wheel-throwing impossible (HULTHÉN 1974, 69). The wheel-thrown pots are built up by hand,

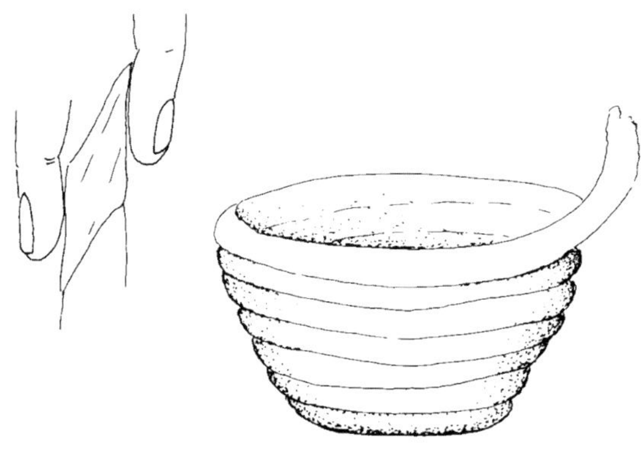

Fig. 46. *The hand-made ceramics at Groß Strömkendorf were built up by coiling and fastened to each others by the so called N-technique (drawing: A. Jeppsson).*

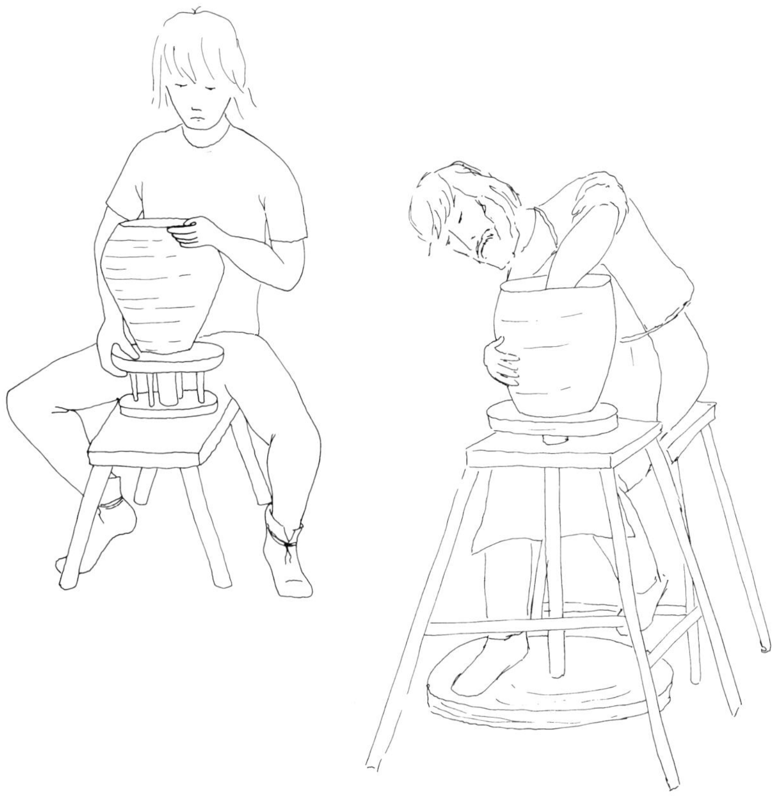

Fig. 47. The major difference between ceramics made on the turntable (left) and the potters-wheel (right) is the speed of the wheel. The turntable was thrown by hand and the potters-wheel by foot (drawings: A. Jeppsson).

with the aid of the centrifugal force. Vessels with traces of being shaped on turntables are normally made by coiling and given a final shape when rotated on the turntable. One of the most important finds connected to the research of the pottery craft was made at an excavation in Dortmund-Groppenbruch in Westphalia (Bergmann 1993). A disc, interpreted as part of a turntable, was found in a layer dated to the 12th century. Unfortunately, no finds of turntables from the medieval period are known in Northern Europe. The best indication that turntables were used for making some of the Slavonic vessels at Groß Strömkendorf are the imprints on six base sherds *(Plate 18,1–2.4)*.

At Groß Strömkendorf the Sukow vessels, which were shaped on turntables, were found in the younger settlement phases and at the cemetery. Only one of these vessels was located in a dendro-dated feature, a well dated to 809. It seems likely that the use of turntables on the Sukow pots at Groß Strömkendorf was connected to the younger period, after 760. The amount of Feldberg pottery shaped on the turntable was much higher than that of Sukow pots. The turntable was an important tool for making distinct decorations, e.g., wavy lines and straight horizontal lines. The oldest dendro-dated feature with finds of Feldberg sherds shaped on a turntable is dated to 762. Sherds of this type furthermore were discovered in all the settlement areas, including a single rim sherd in a pit house in the north-eastern area. Finds from graves also were identified. In spite of the uncertain dating for the first phase at Groß Strömkendorf, it is likely that Feldberg pots were shaped on turntables already around 740.

Surface treatment

All the outer surfaces of the Slavonic vessels were smoothened. There are no finds of pots with further treatment, including burnishing and the so-called "Polierte Ware"[7]. The smoothing was most likely made with the wet hand or a piece of cloth. In general, Sukow vessels have a rougher surface than the other Slavonic vessel types. Feldberg is a group of pottery with smooth as well as rough surfaces. It is possible that the explanation could lie in different chronological phases of the settlement.

Decoration

The most common decoration types on Slavonic ceramics at Groß Strömkendorf are different kinds of wavy lines and/or horizontal lines. Of the 1,981 decorated sherds

7 The "Polierte Ware" is dated to the 8th and 9th centuries and was defined by T. Kempke (1984).

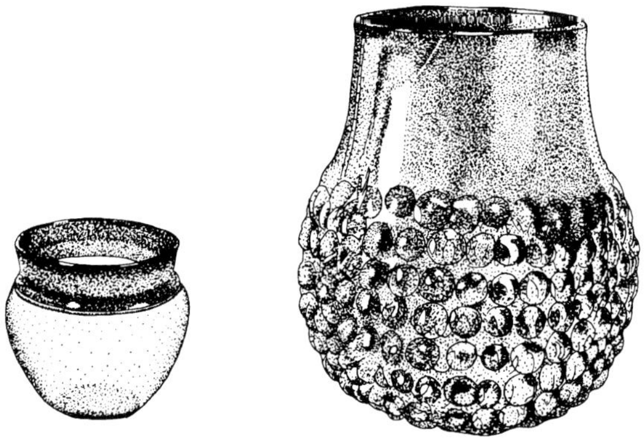

Fig. 48. Beaker of glass. The knobs and the vessel profile could have served as prototype to, e.g., the Feldberg ceramics (after ARWIDSSON *1984, Abb. 24).*

(47.3 kg), 1,194 sherds have wavy lines, 666 sherds horizontal lines and 159 sherds both types. A total of 132 sherds are decorated with different kinds of comb impressions. The earliest finds with this type of decoration are dated from Phase II and onwards. Analyses of nine Slavonic sherds with comb impressions showed that all were made from fine clay (ware groups VI and VII).

The first finds of Slavonic ceramics with cordons most likely appeared during the same period as the comb-decorated pots (*Plate 17,12*). A total of 21 sherds with cordons were identified. Analyses of four of these showed that all of them were made in the same way and belong to ware group VI.

Stamp-decorated pottery was found all over the site, from all settlement phases and from the cemetery. It seems likely that the more simple decoration, the plain circle, appears in each phase. The more complex stamp decorations, e.g., the flower stamp, triangles, and squares, belong mostly to the period after the middle of the 8th century. Only one sherd is related to a pit in the north-eastern area, another one was found in a mound in the cemetery area. The variety of stamps at Groß Strömkendorf is, compared to other Slavonic sites from the 9th and 10th centuries, fairly low. At, for instance, Groß Raden, at least 24 different kinds of stamps were identified (SCHULDT 1981, 33 Abb. 13). It is likely that the contacts with Western Europe and with the Frisians affected the production of Slavonic pottery in Mecklenburg during the 8th century. At least two of the stamp-decorated sherds from Groß Strömkendorf are probably of Frisian origin *(Plate 23,5–6)*. At the excavation, some hundred metres to the east of the trading site, an unusual artefact was discovered. The iron object was broken at the edge but, according to Wietrzichowski, it was most likely used to make stamps on the ceramics (WIETRZICHOWSKI 1993, 34). It was shaped like a rhomb. This type of decoration was identified on a few sherds from the trading site and the cemetery.

Another type of ornaments are plastic decorations, e.g., knobs *(Plate 11,2–3)*. Vessels with knobs mostly belong to Phase II and later. Finds also were made in the north-eastern area and in a boat grave, as well as in two mounds at the cemetery. Apart from the knobs, there are no similarities with the contemporaneous cups found in Scania. This type of vessels may be interpreted as copies of Western European glass beakers decorated with knobs *(Fig. 48)*.

Ceramics with rim decorations mainly were found in the northern, central and southern part of the settlement. Only one of the total material of 120 sherds with rim decoration was found in the north-eastern area. A few sherds were also located in different types of graves. It seems likely that the phenomenon of rim decoration began around 760.

Firing

The Slavonic pottery at Groß Strömkendorf was fired to between 800°C and 900°C, some vessels even at 100°C more. These are rather high temperatures, considering that it is only necessary to fire the clay to temperatures above 450°C. Investigations of Early Slavonic pottery from the settlement at Dessau-Mosigkau in Sachsen showed that the pots were fired to between 520°C and 600°C (KRÜGER 1967, 90). The reason why the ceramics at Groß Strömkendorf were fired to significantly higher temperatures is uncertain. It seems unlikely that all analysed pots were exposed to secondary firing, e.g., a conflagration. The explanation may be that the pots were primarily fired to more than 800°C in either a kiln or an oven. Investigations of the temperatures in an oven (one chamber) showed that the fire could reach more than 1000°C (GOSSELAIN / LIVINGSTONE SMITH 1995, 154 Fig. 5). However, the North European hand-made vessels were normally fired in either bonfires or in pits *(Fig. 49,1–2)*. It cannot be excluded that some of the larger pits or traces of fire on the surface at Groß Strömkendorf were the remains of firing pottery. It is very difficult to identify these types of constructions, and no clear finds have been made in, for example, northern Germany and Scandinavia. A kiln dated to the 9th and 10th centuries was found at Kremitz, Kr. Jessen, in Sachsen-Anhalt (KNORR 1970, 77). The kiln had two separate chambers and was most likely used for firing Feldberg ceramics. The pottery was placed on top of a platform with the fire beneath it *(Fig. 49,3)*. At Järrestad in southern Sweden, another kiln dated to the 8th century was discovered (BRORSSON 2003b, 361 ff.). Similar to the example from Kremitz,

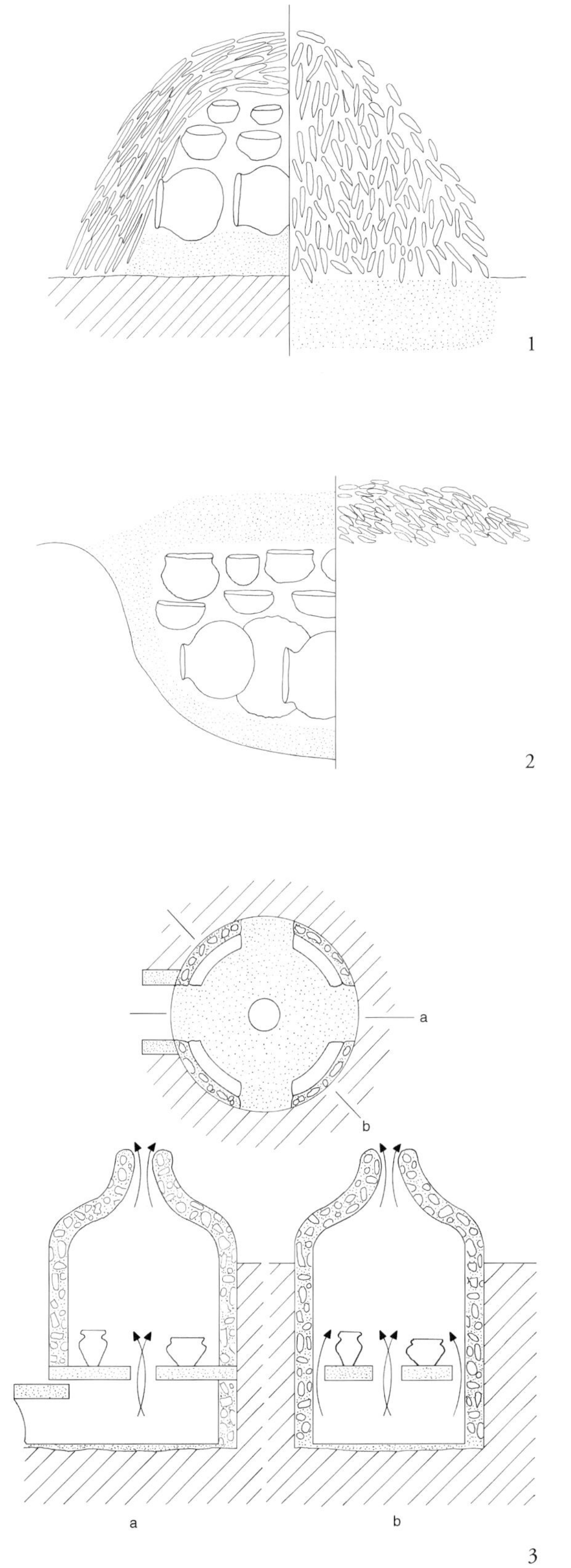

Fig. 49. *Different kinds of firing ceramics. 1 - Bonfire (drawing: A. Jeppsson); 2 - Pit (drawing: A. Jeppsson); 3 - Kiln with two chambers found at Kremitz in Sachsen-Anhalt (Knorr 1970, Abb. 32).*

this kiln also had two chambers. The fire was in a separate chamber at the bottom of the kiln. It was covered with a layer of clay with holes in it. The pots were placed on this layer of clay. The kiln is unique in Sweden and most likely was built by foreign people visiting Järrestad. It cannot be excluded that the Slavonic pottery at Groß Strömkendorf was actually fired in kilns.

Function

The function of the different vessel types has always been a subject of discussion in archaeology. It is normally the size of the pot and the shape of the rim that are studied. A common opinion is that pots with everted rims are well suited as storage containers. It was, for example, possible to stretch and tighten a piece of leather or cloth as a lid on the vessel. The hand-made coarse pots are usually interpreted as cooking vessels, but during the last 15 years other functions, e.g., for storage or as containers during the processes of fermentation, have been suggested (Lindahl 2002b, 39).

The analyses of the hand-made pottery from Groß Strömkendorf showed that the Slavonic, Saxon and Scandinavian pots were undoubtedly used in the household either as cooking or storage vessels. The ceramics from the Rhineland are of a finer ware with thinner vessel walls and were better suited as either storage or table ware.

Lipid and chemical analyses were performed on native Viking Age pottery sherds from four locations in the middle of Sweden. The results showed that the manorial site, Valsgärde, with several rich boat graves, cannot confirm the expected hypothesis (Isaksson 2000, 22). The old Norse sagas tell us about feasts with meat, and the opinion has since been that meat was a very important foodstuff for the Vikings. However, the analyses only found remains of vegetables in the pots from Valsgärde. Other analysed vessels from the hinterland had remains of animal as well as vegetable products.

X-ray and chemical analyses also were made on four pottery sherds from the settlement close to the rampart at Birka. The sherds were from a relief-band amphora, a Slavonic vessel (Fresendorf) dated to the end of the 8th century, a Finno-Ugric vessel, and a local Scandinavian vessel (Gustafsson 1995, 7). These analyses showed that the relief-band amphora may have been used as a container for either wine or beer. It could have had a primary function as a container for wine and a secondary function for beer when it was in use at Birka. The Slavonic pot was probably used as either cooking or table ware. However, based on the shape it seems likely that the vessel was rather used for cooking. The Finno-Ugric pot could have been used as a soup or bouillon bowl and the local vessel as cooking or table ware.

It is indeed important to separate hand-made and wheel-thrown vessels. The wheel-thrown ones were imports obtained as trading goods. The function as storage containers for wine seems plausible. There is no evidence among the pottery from Groß Strömkendorf that the local inhabitants tried to make copies of the imported pots. As mentioned above, the decoration, sherd thickness and quality differ considerably from the local Slavonic pottery.

Another question is why most of the Scandinavian vessels belong to ware group XIII (coarse clay, no added temper) and only a few to group VI (fine clay, crushed rock used as temper). The latter is also the group to which most of the Slavonic sherds belong. The Scandinavian vessels in group VI were perhaps made by people who also made Slavonic pots. However, it could also be the result of contacts with other areas, e.g., East Denmark, where analyses of ceramics from several sites have shown that such ware was typical of this region. One noteworthy detail is that none of the analysed Scandinavian sherds in ware group VI belong to Phase I. They were discovered in dendro-dated wells, in a boat grave and in an inhumation grave, and were dated to Phase II and later. There are three possible interpretations for this phenomenon at Groß Strömkendorf:

- The contacts with Scania began around 760, which is later than the contacts with western Denmark. This hypothesis is supported by the presence of typical Scanian vessels with a protruding foot, dated from the 8th century and onwards.
- The Scandinavian pots were made by Slavonic potters. Similar results were observed in a study of the pottery from Gorodishche outside Novgorod in Russia (Brorsson 2001, 57).
- The pottery was made by Scandinavians who used a technique and raw material also used by the Slavs. It was a result of good relations between Slavs and Scandinavians.

The question is if the different cultural groups at Groß Strömkendorf used each other's pots. It has been suggested that the hand-made pots from the different cultural groups in the Baltic region were not used by other groups in the region (id. 2003d, 232). The main reason is that the vessel, the hearth and the food were correlated to each other, and if something was changed, it affected the others. That could be the reason why the Scandinavians used their own pottery and the Slavonic people used theirs, etc. However, this interpretation may easily be challenged when looking at the artefacts from the cemetery. In the boat graves, for instance, Scandinavian pots were found together with Frisian and Slavonic vessels. It is not likely, according to the grave structure, that these graves were intended for other people than Scandinavians. It must instead have been the result of good relations between different groups so that the origin of the pots was nearly irrelevant.

Chronological results *(Fig. 50)*

The dominating pottery type throughout the 8th century and early part of the 9th century at Groß Strömkendorf is Sukow. There are no indications of any decrease in the amount of this type during the period the trading site existed. The quantity of Sukow pottery is almost constant during the whole period. According to results from other excavations in the Baltic region it is likely that the Sukow ware started to decrease in number during the beginning of the 9th century.

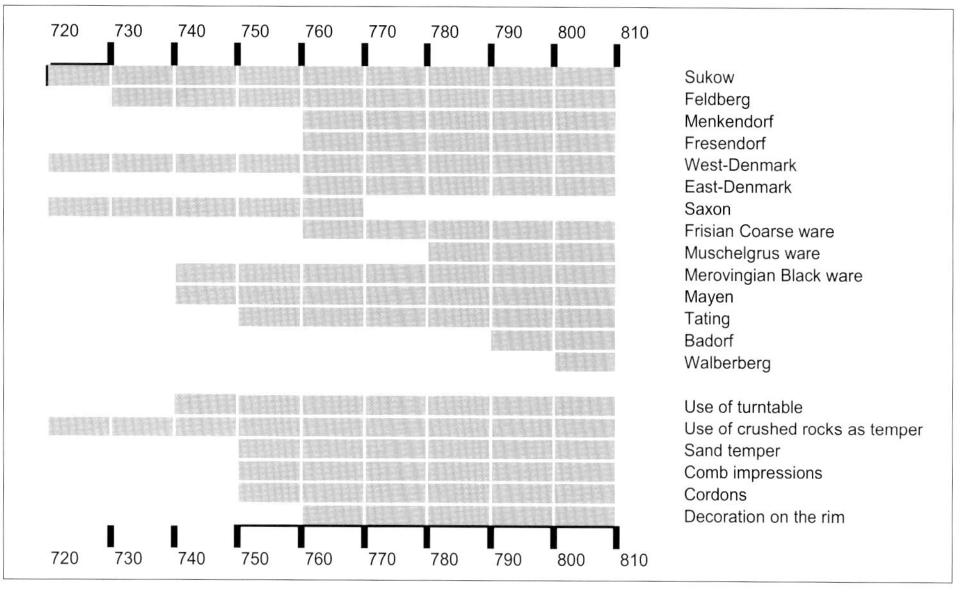

Fig. 50. Chronological distribution of the different pottery types appearing at Groß Strömkendorf. Some datings are based on vague materials and must be considered as preliminary. The connection between the ware, decoration and the Middle Slavonic pottery is clear.

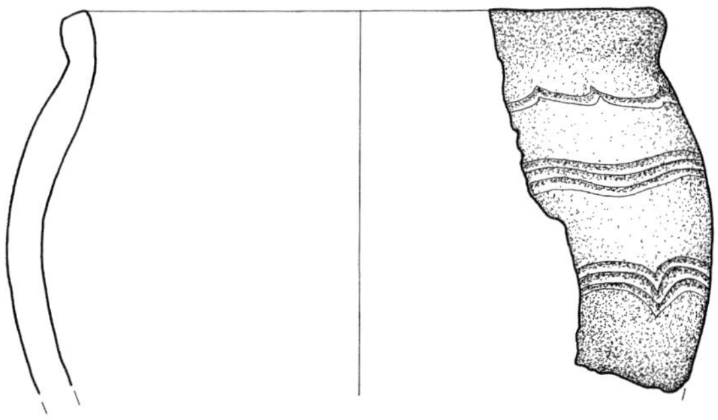

Fig. 51. One of the oldest finds of Feldberg pottery found outside the Slavonic region. The vessel was found in Ribe in a layer dated to 740 (Claus Feveile, pers. comm). Scale 1:2.

The earliest dating of the Feldberg ceramics at Groß Strömkendorf is uncertain. The oldest dendro-dated feature with finds of Feldberg ware is a well dated to after 735. The sherd found in the well has an everted rim with indistinct wavy lines. The vessel was most likely made without the use of a turntable. At the oldest settlement area in the north-east of the trading site, the Feldberg ceramics make up about 25 % of the total Slavonic material. Recent unpublished finds from the "Post Office excavation" in Ribe have shed new light on the earliest Feldberg pottery: a Feldberg vessel with wavy lines was discovered in a dendro-dated layer from the period 720 to 740 *(Fig. 51)* (Claus Feveile, den antikvariske samling, Ribe, pers. comm.). Almost the same dating was confirmed by excavations in Szczecin (Łosiński / Rogosz 1984, 202 Abb. 17).

Different studies of the vessel profiles and the rim shapes strongly indicate a common tradition behind the Sukow and Feldberg ceramics. The main differences between the two types are the presence of decoration on Feldberg and the prevalent use of a turntable for shaping and decorating this type of pottery. However, as noted by, for instance, Wietrzichowski, even undecorated Feldberg vessels existed. An example from Groß Strömkendorf is an undecorated Feldberg pot used as an urn *(Plate 9,13)*. The dating of the grave is uncertain.

The earliest dendro-dated finds of Menkendorf ware at Groß Strömkendorf comprise two different sherds found in the same well (No. 10370-17). The wood in the well was dated to 761/762. The sherds were most likely deposited as waste material when the well was abandoned, which may have been after Phase II (around 780). The geographical parallels to the finds at Groß Strömkendorf are few and the only site with contemporaneous finds is Wolin. The Menkendorf ceramics from Wolin were dated to the second half of the 8th century (Brorsson / Stanislawski 1999, 290 Tab. 1, 2). According to the finds at Groß Strömkendorf it is possible that the first Menkendorf ceramics were produced sometime between 760 and 780. The investigation of the vessel profiles showed that Menkendorf (shape group i) is a new type of vessel. Its appearance could have several reasons. One of them may be the influence of a foreign pottery tradition, e.g., the Carolingian.

The apparently oldest Fresendorf sherd from a dendro-dated context was found in well No. 81-3, situated in the central settlement area. The well was dated to 779. Fresendorf sherds also were found in the northern settlement area, dated from 760 to 780. In a concentration of pits in the north-eastern part of the settlement, sherds from two Fresendorf *(Plate 15,1.3)* and five Feldberg *(Plate 11,3.6.8)* vessels were found. A base sherd, which was classified as Slavonic, with imprints from a turntable was also recovered from one of the pits (No. 931-17) *(Plate 18,4)*. The Fresendorf pottery at Groß Strömkendorf has rather good parallels in the ceramics from Wolin, dated to the second half of the 8th century (Brorsson / Stanislawski 1999, 291 Tab. 2). The first ceramics of Fresendorf type at Groß Strömkendorf can be dated to after 761/762 and before 779. It is important to note that the Fresendorf type, among other attributes, has a new rim type. This could be the result of contacts with other people. Generally, this vessel type was concentrated on Rügen and could have been associated with the Slavonic tribe of the Ranes *(Fig. 15)*.

The finds of uncertain Ladoga *(Plate 26,3)* and Curonian *(Plate 26,4)* vessels at Groß Strömkendorf illustrate some of the problems when dealing with pottery. Ladoga vessels were the dominating vessel type used by the East Slavs. It was strongly influenced by the Menkendorf ware (Nosov 1992, 55). However, the Ladoga vessels were coil-built and not shaped on turntables. The date for the Ladoga ceramics from other excavations is mainly the period between the 9th and 11th centuries. However, the vessel at Groß Strömkendorf is more or less identical with the Ladoga type, both in shape and forming technique. The ware differs from the large group of Slavonic pottery at Groß Strömkendorf and belongs to ware group XI (medium-coarse clay with crushed granite used as temper). Ladoga vessels with the same type of ware were found at Gorodishche (Brorsson 2001, 56 Tab. 1). However, the dating of the vessel from Groß

Strömkendorf to the 8th century makes it highly unlikely that the vessel is a genuine Ladoga vessel.

The vessel with finger imprints on the rim interpreted as a Curonian pot *(Plate 26,4)* belongs to the largest ware group, No. VI (fine clay with crushed rock used as temper).It includes most of the Slavonic vessels as well. The vessel may have originated from present-day Latvia, but could also have been produced in the hinterland of Groß Strömkendorf. However, the typological attributes have more in common with the Curonian pottery tradition than with the West Slavonic. It can also be noted that most of the local Curonian vessels from Grobina were made in the same way as the pot from Groß Strömkendorf (Brorsson 2003c, 368).

Quite unexpected in this early context are sherds from bowls. One of the sherds was discovered rather deep in a well from the southern settlement area. The well was dendro-dated to 806. Other finds from this well are Sukow, Feldberg, Muschelgrus ware and Badorf sherds. Another sherd of a bowl is a surface find. The Slavonic bowls are generally classified as Garz type, with a much younger dating. The oldest finds of Garz bowls in, for instance, Szczecin, derive from the middle of the 12th century (Łosiński / Rogosz 1984, 202 Abb. 17). At the excavations in Starigard/Oldenburg, Garz bowls were discovered in layers dating mainly to the 11th and 12th centuries (Kempke 1984, 90 Tab. 1). This type of bowl also appears in the Baltic ware from the third quarter of the 11th century. Thus, the finds at Groß Strömkendorf are much earlier than at other known places.

The Scandinavian influence at Groß Strömkendorf has been confirmed for all the different phases, including the cemetery. It seems likely that the Scandinavians were an important part of the population at the trading site. The study of the faunal remains has shown that a fairly large amount of cod was among the excavated material (Schmölcke 2004, 100 ff.). Cod is usually rare in the Slavonic area during medieval times, but is characteristic of the Scandinavian regions.

The contact with Scandinavia seems to have been primarily concentrated on western Denmark, where Ribe was most likely one important partner in the trading network in Northern Europe. The contacts with eastern Denmark (Scania) were not established before about the third quarter of the 8th century, the same period in which the first Slavonic pottery appeared in Scania (Brorsson 2003d, 232).

The Saxon influence at the site is only noted in the early phase of the settlement and at the cemetery.

The first finds of Frisian Coarse pottery were identified in the northern settlement area, dated to between 760 and 780. No ceramics of this type were found in the dendro-dated wells. The Frisian influence among the ceramics at the cemetery is extremely low; the only find of Frisian Coarse ware is from boat grave No. 206 *(Plate 23,1)*. The Muschelgrus ware derives from Phase III in the southern area, which is dated to between 780 and 811.

The Vorgebirge-Eifel ceramics appear already at the north-eastern settlement, dated to the first half of the 8th century. Two Mayen sherds were found in a pit house. The earliest dendro-dated find is a sherd of Merovingian Black ware from 782. The amount of sherds from Vorgebirge-Eifel is almost the same in the northern and southern settlements areas. Only limited material is related to the graves, and the finds are partly concentrated on one area. The sherds were mainly discovered in mounds and in urn graves. The oldest find of Badorf pottery derives from a feature dendro-dated to 806, which fits well with the finds in the southern area. The Tating ware was found in the northern settlement area and it seems likely that the pottery appeared between 760 and 780 at Groß Strömkendorf. No sherd of Tating was located in any of the dendro-dated features.

The dating of the two finds of relief-band amphorae is quite uncertain. The central area, where the finds were made, is dated to a long period, in particular to the 8th century. However, the waste material in the pits seems also to have been deposited during the beginning of the 9th century.

The thermal analyses as well as the microscopic analyses have shown that there are at least two productions of Tating ware and three types of Mayen ware. The six analysed Badorf vessels are divided into three different productions. This result is not surprising and illustrates the variety in the selection of raw material, as well as the large amount of productions present in the Vorgebirge-Eifel region *(Fig. 18)*. The origin of the Tating productions is still uncertain, and the connection to Mayen can neither be confirmed nor refuted. One of the aspects of the classification of the Walberberg pottery is that the largest grains are between 0.4 and 0.6 mm. The detailed recording of this type of pottery at Groß Strömkendorf confirmed the classification system. The mean size of the measured grains in the sherds is 0.6 mm.

The technological studies of especially the Slavonic ware have shown a homogenous pottery craft. Only three vessels, one Sukow, one Feldberg and one Menkendorf, differ from the rest of the material in that they were made from sand-tempered clays. The sherds were found in different pits in the central area and their age is uncertain. One of the pits (No. 166-3) also contained sherds of Muschelgrus ware, Badorf and relief-band amphorae. It seems likely that at least this pit dates to the youngest phase, perhaps to the beginning of the 9th century. It has been suggested that the decoration of the Slavonic ceramics was influenced by the Carolingian Empire. It is not unlikely that the use of sand temper was also a western influence.

Groß Strömkendorf as a trading place in Northern Europe

The large pottery inventory from Groß Strömkendorf has been subject to various analyses. The typological and chronological studies on the pottery were intended to either support or refute other investigations of the trading site, such as the important work by A. Tummuscheit (2003a) on the structures of the settlement. This pottery study has demonstrated several aspects that support hypotheses concerning dating and distribution of the settlement, and the connection between cemetery and settlement has been revealed.

It seems likely that the buried people at the cemetery were mainly of Slavonic descent. Urns of especially Sukow or Feldberg type were commonly used in the burials. Also in the mounds these types are frequent. The graves with Slavonic pottery are distributed all over the cemetery. One of the most unique graves is urn grave No. 800. The pot used as urn is a Merovingian Black ware *(Plate 24,1)*, and it may, according to the burial custom, be interpreted as a Slavonic grave. This grave was perhaps designed for an especially important person, who may have received the pot as a gift or bought it through a trading alliance. The vessel could have been in the family for a decade and was finally used as the final resting place for the owner. Several sherds of Vorgebirge-Eifel origin were identified in different graves in the western part of the cemetery. The question is if the sherds were mixed with the older settlement material or were deliberately put into the graves. The sherds in that specific area consist of Merovingian Black ware, Mayen, Tating and Walberberg. At least the latter sherds are most likely from the period after the settlement and may be considered grave goods. The small sherds of Vorgebirge-Eifel origin were perhaps put in the graves as *pars pro toto* (the part of the whole), representing complete vessels. Maybe it was not the complete vessel that was of importance, but a piece of the artefact. This type of find also was discovered at, for instance, the Brudager cemetery on Funen in Denmark (Stilborg 1997, 193).

The Slavonic pottery from the trading site and the cemetery is more or less of the same ware type and vessel shape. It is therefore likely that the pots at the cemetery were produced at the same place as the pots found at the trading site. Also, some burial vessels may previously have been used in the household. This result has, for example, been confirmed at the Early Iron Age site of Möllegårdsmarken on Funen in Denmark (ibid. 307 f.). In some of these graves, fine-ware vessels and sherds of coarse-ware vessels were found. The latter could have been put into the graves to provide the dead person with a functional set of pottery for the afterlife. Reused vessels, which were put in graves, have also been found in other societies[8].

The boat graves with different kinds of pottery are also an interesting phenomenon. The burial tradition may be interpreted as Scandinavian. However, several other vessel types, besides the typical Scandinavian ones, were found in the graves. The same pattern is seen in the inhumation graves. Two main conclusions may be drawn: Since there are examples of Slavonic pots in boat graves as well as pots of Merovingian Black ware in urn graves, it is difficult, if not impossible, to use the vessels to point out the exact cultural affiliation of the dead person. Secondly, it seems likely that several different groups lived quite integrated. The origin of the vessel was probably of minor importance. The only coffin grave, which was most likely Christian, contained a Mayen sherd together with five sherds classified as Slavonic. A possible chamber grave also was discovered (Gerds in prep.), containing four coarse hand-made sherds, for which no further classification was possible. From the 40 sherds found in different animal graves, such as horse or dog graves, only six sherds were classified. A single sherd in a horse grave is of Scandinavian origin and the rest are either Sukow or just Slavonic. These examples, together with the structures at the cemetery, indicate that at least Slavonic and Scandinavian people were buried at Groß Strömkendorf.

Several similarities between Groß Strömkendorf and other sites in the Baltic region have been established. The most important contemporaneous sites, which have been thoroughly investigated, are Wolin and Ribe. Analyses of ceramic thin sections were made on material from the 9th and 10th centuries in Wolin (Brorsson / Stanisławski 1999, 294 Tab. 4). The analyses showed that most of the pottery was made of sorted fine clays tempered with

8 Salskov Roberts 2002, 28. Some examples are from Delphi and Keos in Greece.

crushed granite. There were no differences between the Feldberg and Fresendorf wares. As a complement to these published analyses, an additional three Sukow sherds were included in that study. The sherds were found in the earliest layers, dated to end of the 8th and beginning of the 9th century. All of them were made in the same way as in Groß Strömkendorf. The Sukow pots from Wolin were made from the same type of clay and granite as the vessels from Groß Strömkendorf.

The pottery from the Danish trading site of Ribe also exhibit similarities with Groß Strömkendorf. The excavation at the Post Office in Ribe showed that the composition of different hand-made and wheel-thrown pots was in several aspects the same as in Groß Strömkendorf. Besides the dominating finds of Slavonic and Scandinavian sherds, the rest consists of nearly the same types. The Muschelgrus ware and the Merovingian Black ware appeared at about the same time. The most common pottery type found in Ribe is the so-called Scandinavian "halvkuglekar" (semi-bowl vessel). This type also was found at Groß Strömkendorf. In Ribe, a quite early find of Slavonic Feldberg pottery was made *(Fig. 51)*, which is common at Groß Strömkendorf as well. An important connection between the Slavonic and the Danish trading sites is the quality of the Scandinavian ware. The Scandinavian ceramics at Groß Strömkendorf belong mostly to ware group XI, which is coarse clay without any added temper. Local Scandinavian ceramics from Ribe were, according to macroscopical investigations, made in the same way. This observation has been supplemented by thin section analyses of two local Scandinavian sherds from Ribe, dated to between 725–760 and 760–780 respectively. The sherds were found at the Post Office excavation. Both sherds were made of coarse clays without any added temper. These sherds are of the same character as most Scandinavian sherds at Groß Strömkendorf.

Another important connection between Groß Strömkendorf and Ribe was drawn by A. Pöche in his study of the beads from Groß Strömkendorf. Beads of Ribe-type were found in the same area of the central settlement as a concentration of Scandinavian sherds (Pöche 2005, 99 Abb. 38). In this area, a large amount of cod bones was also found (Schmölcke 2004, 31 Abb. 7). As mentioned earlier, cod is a fish that the Scandinavian population ate, but was rejected by the Slavs.

It would have been interesting to compare the pottery from Groß Strömkendorf with more contemporaneous sites from the southern Baltic countries than just Ribe and Wolin. However, only a few sites with fairly large amounts of ceramics have been excavated, and a thorough investigation of the pottery is not the rule.

Other studies have shown that Groß Strömkendorf belonged to some kind of trading network in the south-western Baltic Sea area (Müller-Wille 2001). Ribe, as an early urban centre, was most likely one of the most important sites for trading connections in the northern sphere.

The relation to Scania, especially Åhus and Löddeköpinge, seems to have been of minor importance than that to Ribe. Analyses of local Scandinavian ceramics from six locations in Scania showed that the potters preferred fine or medium-coarse clays and tempered them with crushed granite. About half of the Scandinavian material found at Groß Strömkendorf was made from these types of clays. As mentioned above, this could be the result of the first contacts with Scania after around 760. It corresponds fairly well with the dating of Slavonic ceramics of Feldberg and Fresendorf types in Scania. The Slavonic influence at the Scanian trading sites is dated to the period between 775 and 825 (Brorsson 2003d, 232), and this result is mainly based on stratigraphy and ceramics from dendro-dated contexts from Germany and Poland.

Recent analyses of ceramics from Kaupang have yielded the same results as for the Groß Strömkendorf pottery. Analyses of semi-bowl vessels showed that the ceramics were made from naturally coarse clays without any added temper (id. 2003a). The vessels from Kosel, Schleswig-Holstein, were made the same way (id. 2006). Another important conclusion is that the Muschelgrus ware found at Kaupang and at Groß Strömkendorf was most likely produced in the same region.

Based on the grave material, the Scandinavian and Saxon influence has been interpreted as representing people from these areas visiting the trading site.

A question that must be considered is how the Vorgebirge-Eifel and the Frisian ceramics reached Groß Strömkendorf. The wheel-made vessels have, according to analyses of sherds from Birka, been interpreted as containers for wine. It seems likely that these types of vessels came to Groß Strömkendorf and other trading sites in Northern Europe through places like Dorestad. The vessels do not represent people coming from the Vorgebirge-Eifel region to Groß Strömkendorf; rather, the vessels or their contents must be seen as a trade goods. The Frisians have been considered one of the most important trading communities in Northern Europe. Their vessels were, similar to the Slavonic, hand-made and most likely not the proper trade goods. Perhaps it was the content that was important, but since the Frisians were a travelling people, the ceramics may also represent traders visiting Groß Strömkendorf.

A remarkable result is the distribution of the Scandinavian and Vorgebirge-Eifel ceramics in the central settlement area. Does this distribution indicate that wheel-thrown pottery was mainly used by the Scandinavians, or have other aspects affected the spread of the sherds? It is very likely that the Scandinavians, together with the Frisians, were one of the main intermediaries in the contacts between Groß Strömkendorf and Western Europe. They may have brought the Vorgebirge-Eifel pottery to Groß Strömkendorf from western Denmark or perhaps even from as far away as Dorestad. A natural trading route a few miles north of the modern Kiel Canal existed already during the Early Medieval period (Sawyer 1971, 186). The route passed Hollingstedt, Haithabu and Schleswig (Ellmers 1972, 229 Karte 1).

From Denmark, the traders could have travelled to Mecklenburg and further to the east and the north, e.g., Wolin and Scania. According to the hypothetic map of contacts in the Baltic region made by Callmer, Groß Strömkendorf/*Reric* in Mecklenburg could have been in direct contact with western Denmark and in the east with places like Rostock-Dierkow (Callmer 1992, 102 Fig. 2). There were no direct contacts between Groß Strömkendorf and Scania. Instead, the trade probably passed through other minor trading sites in the region. The connections and the social relations to especially Rostock-Dierkow is one major research task for the future. Sukow, Feldberg and Menkendorf pottery as well as a few sherds of Muschelgrus ware, Badorf and Tating were identified at Rostock-Dierkow, dendro-dated to between 760 and 817 (Warnke 1992).

Another important prospect is to examine the hinterland of Groß Strömkendorf. So far, this only has been carried out on a limited scale, and thus the knowledge concerning the interaction between the trading site and its surroundings is limited (Schmitz 1999; Müller-Wille / Tummuscheit 2004, 37). The study of the handmade ceramics from Hedeby shows that the local material is dominated by "Kugeltopf" and "Eitopf" (Hahn 1977, Abb. 19), two vessel types that were common in the hinterland of Hedeby. The vessel material from the hinterland of Groß Strömkendorf in the 8th century was in all likelihood dominated by Sukow and Feldberg pots. The local pottery and the cultural behaviour of the people from the hinterland most likely played an important role in the daily life at the trading site.

General results

The study of the pottery from Groß Strömkendorf has proved that it is possible to use ceramics not only for chronological interpretations but also for other important conclusions. It has shown the importance to approach this special find category from different angles and with different methods. With regard to the results of the investigation it has been possible to answer the questions raised in the second chapter:

1. The settlement structure has previously been studied by Tummuscheit (2003a). She identified three different phases into which the ceramic material was divided. The pottery study has shown that it is possible to distinguish an older and a younger phase. In features of the oldest settlement phase, Slavonic pottery of Sukow and Feldberg type predominates. The imported material is limited and consists mostly of Scandinavian sherds. The material related to the youngest phase comprises several different Slavonic vessel types. The imported vessels are of Scandinavian types but also of West European origin. Muschelgrus ware only was found in features dated to this youngest phase of the settlement.

The Scandinavian material was found all over the settlement, but a majority of the sherds derive from a section of the central area, which was probably once inhabited by Scandinavians. This interpretation is in agreement with the studies of the faunal remains and the beads.

2. It seems likely that at least Slavonic and Scandinavian people were buried at the cemetery. Among them might have been traders, but the pottery study does not offer tangible clues in this regard. The cemetery was probably also used by others, e.g., people from the Slavonic hinterland. However, the fact that vessels of Frisian and Saxon origin were used as grave goods cannot be interpreted as indicative for the same origin of pots and people.

3. The same types of pottery were found at the settlement and at the cemetery. Thus, hand-made vessels with the same shape, decoration and ware were used both as household vessels and as grave goods. Some of the vessels may have had a primary function in the household and were later put into graves. There are no indications of any vessels primarily purposed for a burial. Yet, the study of the vessel height has shown that the vessels for the graves were more homogenous in size than those located at the settlement.

The wheel-made vessels from the Vorgebirge-Eifel area were mostly found at the settlement. Only one vessel obviously was used as an urn.

4. The amount of imported pots at Groß Strömkendorf is fairly high: Slightly less than 10 % of the classified sherds are of non-West Slavonic origin. The imported pottery originated from Scandinavia, Saxony, Frisia and the Vorgebirge-Eifel area, which includes such important sites as Mayen and Cologne. One vessel may be from Latvia and another is similar to Ladoga vessels originating from present-day Russia. Indeed, the pottery study supports the interpretation that the Groß Strömkendorf was a trading site within a North European trading network during the 8th and early part of the 9th century.

5. The Slavonic pottery at Groß Strömkendorf comprises Sukow, Feldberg, Fresendorf and Menkendorf types. Minor pieces of lids and bowls were identified. The study of the decoration and the shapes of the rims and vessels, as well as the analyses of the firing temperature and the ware have resulted in some important conclusions. Different vessel types display the same ware. A development toward a finer ware from Sukow up to Menkendorf is not discernable. The different Slavonic vessels types were all made by the same type of clay and temper material. However, minor differences in the sherd thickness were established. The Sukow vessels are slightly thicker than the rest of the Slavonic vessels. This may be the result of the use of a turntable for shaping and decorating the Middle Slavonic pots.

A study of the rim shapes has shown a connection between the Feldberg and the Sukow pots. Most of the Feldberg vessels were made in the same manner as the Sukow. Some kind of development towards more distinct rims seems to have taken place. Before the appearance of the Fresendorf and Menkendorf types, there are no major changes in the shape of the vessels.

The study of the decoration has shown that the most significant change occurred between the Sukow and Feldberg vessels. The Sukow ware was undecorated, whereas the Feldberg was the first Slavonic vessel type with decoration.

Sand temper among the Slavonic material was found in sherds dated to the youngest phase of the settlement. This could be the result of influences from Western Europe.

6. The pottery can also be used as part of the interpretation of Groß Strömkendorf being *Reric*, a site mentioned in the written sources. The composition of the assemblage, including Slavonic and imported vessels from several regions in Northern Europe, indicates that the site was most likely an important trading site. The dating of the pottery is of course supported by other studies, e.g., the dendrochronology of wooden features. But even without these important dates, the ceramics would at any rate point to a broad dating to the 8th and early part of the 9th century. The strong domination of the Sukow type suggests that a dating to the 8th century is most probable, while the appearance of Menkendorf, Tating, Badorf and Muschelgrus ware dates more closely to around 800. The dating of the pottery, together with the composition of the different vessel types, indicates that the site was in use during the same period as *Reric*. Thus, the pottery study gives no reason to refute the possibility that Groß Strömkendorf is the ancient trading site *Reric*.

Summary

More than 62,000 pottery sherds were found during excavations at Groß Strömkendorf between 1990 and 1995. The material was thoroughly documented, and several different analyses were performed. The documentation contains information about sherd thickness, type and amount of temper, vessel type, decoration, size and shape of the rim and base. Microscopy of thin sections, grain size analyses, thermal analyses and shape analyses were performed on a sample of sherds.

The excavated assemblage represents different settlement phases, as well as a large amount of material from the cemetery. It was possible to reconstruct 100 vessels from the rim to the base. These vessels were mainly found as urns in graves but a fairly large amount of pots also derives from wells, pits and pit houses.

The Slavonic vessels dominate the assemblage with more than 90 % of the total material. Of the Slavonic vessels, the Sukow type is the most common with about 70 %. Feldberg vessels are the second most common Slavonic vessel type, followed by Menkendorf and Fresendorf. Bowls and lids also were identified. The analyses and the detailed studies of the ware have shown that there are nearly no differences between the discrete Slavonic types. There are some indications that thinner walls as well as sand tempering appeared from the late 8th century or beginning of the 9th century onwards. Thinner vessels could be the result of the use of the turntable, and sand tempering could imply connections with the West European pottery craft.

The pottery from the cemetery and the trading place at Groß Strömkendorf has shed a new light on the investigation of the contacts between different regions in the Baltic area. Several significant results have been achieved thanks to the ceramics, the dendro datings, the settlement patterns as well as the distribution of the different kinds of graves. The site was most likely founded during the early part of the 8th century by Slavs, Scandinavians and Saxons.

The question about the beginning of the Viking Age is also interesting. As mentioned above, different artefacts or historical events have been used to determine the incipient stage of the Viking Age. The trading site at Groß Strömkendorf and its ceramic assemblage can add to this discussion. The pottery from the 8th century is identical to the ceramics in Scandinavia between the 8th and 10th centuries – no changes in either ware or vessel shapes appear. Those three centuries are homogenous and, regarding the ceramics, the Viking Age may already have commenced at the beginning of the 8th century. The Scandinavian influence at Groß Strömkendorf during the same time indicates that the people began with trade or raids more than 70 years before Lindesfarne.

It seems plausible that the traders were among those buried at the cemetery at Groß Strömkendorf. The pottery and the burial customs indicate that people from several regions in Northern Europe were buried there. However, it cannot be excluded that the cemetery was also intended for a rural Slavonic settlement nearby the trading site.

One of the main questions is if Groß Strömkendorf is identical with the site of *Reric* from written sources. This is of course a question that has to be answered by looking at several aspects of the site. The dendro datings in relation to the rich imported artefacts definitely confirm that this is a credible conclusion. The imported ceramics from especially the Vorgebirge-Eifel region and the North Sea coast can be taken as strong evidence, but it is also important to bear in mind that finds of these vessel types have been made on other places in the Baltic region. Additionally, even without the dendro datings, the pottery would have been most likely dated to the 8th and early part of the 9th century.

The study of the ceramics from Groß Strömkendorf has shown that it is important to look at a large pottery material with different methods, e.g. thin section analyses, but that the archaeological interpretation is still the most important. In this case, no analyses can replace the archaeologist's hands.

Zusammenfassung

Die Keramik des frühmittelalterlichen Handelsplatzes und Gräberfeldes von Groß Strömkendorf bei Wismar, Mecklenburg

Zwischen 1990 und 1995 wurden während der Ausgrabungen bei Groß Strömkendorf mehr als 62.000 Gefäßscherben geborgen. Das Fundmaterial ist sorgfältig dokumentiert und aufgenommen worden. Es wurde mehreren unterschiedlichen Analysen unterzogen. Die Fundaufnahme bezog sich auf die Wandstärke der Scherben, Art und Quantität der Magerung, Bestimmung des Gefäßtyps, Verzierung sowie Durchmesser und Form des Randes und des Bodens. Bei den an einer Auswahl von Scherben durchgeführten Analysen handelt es sich um mikroskopische Untersuchungen von Dünnschliffen, der Bestimmung von Korngrößen, um thermische Analysen und die Bestimmung der Formen.

Das ausgegrabene Fundmaterial repräsentiert mehrere Siedlungsphasen. Es stammt außerdem in großer Zahl von dem Bestattungsplatz. Insgesamt gelang es, 100 Gefäße oder Gefäßprofile vom Rand bis zum Boden zu rekonstruieren. Die vollständig erhaltenen Behältnisse sind vor allem als Urnen in Gräbern gefunden worden. Ebenfalls stammt eine recht große Zahl an Gefäßen aus Brunnen, Gruben und Grubenhäusern.

Slawische Gefäße dominieren das keramische Fundgut mit mehr als 90 %. Von diesen ist der Typ Sukow mit etwa 70 % am häufigsten vertreten. Gefäße vom Typ Feldberg stehen zahlenmäßig an zweiter Stelle. Als weitere vorkommende slawische Keramikgruppen sind die Typen Menkendorf und Fresendorf zu nennen. Schalen und Deckel wurden ebenfalls als solche identifiziert. Die Analysen und detaillierten Untersuchungen der Keramik haben ergeben, dass kaum Unterschiede zwischen den slawischen Typengruppen zu verzeichnen sind. Einige Beobachtungen deuten darauf hin, dass dünne Gefäßwandungen sowie Sandmagerung vom späten 8. Jahrhundert oder vom frühen 9. Jahrhundert an vorkommen. Dünne Gefäßwandungen werden als Resultat des Gebrauchs der Drehscheibe gewertet, während Sandmagerung als Beleg von Verbindungen zum westeuropäischen Töpferhandwerk angesehen wird.

Die Keramik sowohl von dem Bestattungs- als auch Handelsplatz Groß Strömkendorf hat neues Licht in die Untersuchungen der Kontakte zwischen unterschiedlichen Regionen im Ostseeraum gebracht. Mehrere wichtige Ergebnisse sind dank der Untersuchungen der Keramik, der dendrochronologischen Daten, des Siedlungsmusters sowie der räumlichen Verteilung der verschiedenen Bestattungsformen gewonnen worden. Mit großer Sicherheit ist der Platz im frühen 8. Jahrhundert von Slawen, Skandinaviern und Sachsen gegründet worden.

Die Frage nach dem Beginn der Wikingerzeit ist von besonderem Interesse. Wie oben erwähnt, sind unterschiedliche Fundgruppen und historische Ereignisse zur Datierung des Beginns der Wikingerzeit herangezogen worden. Der Handelsplatz von Groß Strömkendorf als Ganzes und das keramische Material im Besonderen liefern in diesem Zusammenhang weitere Daten. Die Keramik des 8. Jahrhunderts ist mit der des 8. bis 10. Jahrhunderts in Skandinavien identisch. Es lassen sich weder Veränderungen bezüglich der Warenart noch der Gefäßformen feststellen. Diese drei Jahrhunderte verlaufen, was die Keramik betrifft, homogen und die Wikingerzeit mag bereits zu Beginn des 8. Jahrhunderts begonnen haben. Der skandinavische Einfluss in Groß Strömkendorf während dieser Zeit zeigt an, dass überregionaler Handel oder auch Raubzüge bereits 70 Jahre vor dem Überfall auf Lindesfarne stattfanden.

Auch Händler dürften auf dem Bestattungsplatz von Groß Strömkendorf beigesetzt worden sein. Sowohl die Keramik als auch die Bestattungssitten lassen erkennen, dass Menschen aus unterschiedlichen Regionen Nordeuropas hier ihre letzte Ruhe fanden. Allerdings kann nicht ausgeschlossen werden, dass die Nekropole den Bewohnern einer ländlichen slawischen Siedlung, die sich in der Nähe des Handelsplatzes befand, ebenfalls als Friedhof diente.

Ob mit dem Handelsplatz von Groß Strömkendorf das alte Reric entdeckt worden ist, stellt eine der wichtigsten Fragen dar und kann nur unter Berücksichtigung einer ganzen Reihe von Aspekten beantwortet werden. Betrachtet man die Dendrodaten im Verhältnis zur Datierung der qualitätvollen importierten Fundstücke, so wird diese Vermutung gestützt. Die importierte Keramik, die insbesondere aus dem Rheinischen Vorgebirge und der Eifel sowie der Küstenregion der Nordsee stammt, lässt ebenfalls kaum einen anderen Schluss zu. Allerdings sollte nicht vergessen werden, dass Funde solcher Gefäße auch anderswo im Ostseeraum registriert worden sind. In jedem Falle wäre die Keramik auch ohne die gewonnenen Dendrodaten in das 8. und frühe 9. Jahrhundert datiert worden.

Die Untersuchung der Keramik aus Groß Strömkendorf hat gezeigt, wie wichtig es ist, ein solch umfangreiches keramisches Fundmaterial mit verschiedenen Methoden, wie zum Beispiel Dünnschliffen, zu betrachten. Dennoch ist die archäologische Interpretation immer noch unverzichtbar – des Archäologen Hand kann durch keine wissenschaftlichen Analysen ersetzt werden.

Translation by Dietrich Meier

Резюме

Керамика с торгового пункта и некрополя у Гросс Штрёмкендорф (Висмар, Мекленбург)

Во время археологических раскопок 1990-1995 гг. около деревни Гросс Штрёмкендорфа были обнаружены более 62.000 фрагментов керамики. Весь обнаруженный материал был тщательно задокументирован, и подвергнут различным анализам. Документация материала включала регистрацию следующих параметров: толщина стенок, состав глины, количество примесей, тип сосуда, орнамент, диаметр и форма венчика и днища. В анализы выборки сосудов входили: микроскопический анализ черепков, определение размеров и форм примесей, и термический анализ.

Обнаруженный археологический материал относится к нескольким хронологическим фазам поселения. Большое количество материала было найдено на территории некрополя. Всего удалось восстановить 100 сосудов или профилей от венчика до дна. Целые сосуды, чаще всего находят в погребениях, которые использовались в качестве урн. Так же много сосудов выявлено в колодцах, хозяйственных ямах и углубленных жилищах.

Славянские формы сосудов составляют более 90% из общего числа найденной керамики. Чаще всего встречается тип Суков, насчитывающий около 70%. Второе место по количеству находок занимает тип Фельдберг. В числе славянской керамики были определены также типы Менкендорф и Фрезендорф. Среди известных форм следует упомянуть чаши и крышки. Анализ и детальное исследование керамики позволили установить отсутствие различий между группами славянской керамики. Некоторые наблюдения указывают на то, что тонкие стенки и наличие песка в тесте появляются с конца VIII-го – начале IX-го века. Тонкие стенки сосудов определяются также как индикатор использования гончарного круга, в то время как наличие песка в тесте – как последствие связей с традициями гончарных мастерских Западной Европы.

Керамика из Гросс Штрёмкендорфа вносит новые интересные материалы для изучения связей между различными регионами побережья Балтийского моря. Много выводов предоставляет анализ дендрохронологических данных, информация о развитии поселений и пространственного распределения различных форм погребений. Появление поселения у Гросс Штрёмкендорфа, в основании которого участвовали славяне, скандинавы и саксы, с большой уверенностью можно отнести к началу VIII-го века.

Особый интерес представляет вопрос о начале Викингского периода. Как отмечено выше, для определения начала Викингского периода были использованы различные категории находок, а также исторические события. Торговый пункт у Гросс Штрёмкендорфа предоставляет новые данные для решения этого вопроса. Керамика этого поселения VIII-го века идентична керамике VIII-X вв. из Скандинавии. Отсутствуют и различия в составе теста и формах найденной керамики. Развитие керамики этих трёх столетий происходит равномерно, таким образом, начало Викингского периода можно отнести к раннему VIII-му веку. Скандинавское влияние на поселении у Гросс Штрёмкендорфа указывает на то что межрегиональная торговля или же набеги начались за 70 лет до нападения на Линдесфарн.

В некрополе Гросс Штрёмкендорфа, известны, помимо прочих, также погребения торговцев. И керамика, и погребальный обряд указывают на то, что в этом могильнике захоронены представители различных регионов северной Европы. Нельзя исключить и возможность использования некрополя, как такового, жителями близлежащего славянского поселения.

Одним из важнейших вопросов исследования является проблема возможности идентификации торгового пункта у Гросс Штрёмкендорфа с древним Рериком. В поддержку этого предположения могли бы быть использованы сравнения дендрохронологических данных с датировкой определенных находок импортного происхождения. Находки импортной керамики, особенно из Рейнского предгорья, региона Ейфеля и побережья Северного моря поддерживают эту гипотезу. Однако, нельзя забывать о том, что находки данных типов сосудов известны и на многих других поселениях и могильниках побережья Балтийского моря. С другой точки зрения интересно заметить, что данную керамику, возможно, датировать VIII–началом IX-го века и в отсутствии дендрохронологических данных.

Исследование керамики с торгового пункта у Гросс Штрёмкендорфа указывает на важность использования различных методов как, например метод микроскопического анализа черепков. Тем не менее, археологическая интерпретация материала все еще сохраняет свою необходимость в «руке археолога», которую не могут заменить никакие научные анализы.

Translation by Loretta Popa

Area	Feature	Type of feature	Variety of tree	No. of growth rings	Beginning	End	Fell year	Building year
North-eastern	473-3	Well	Ash	75	661	736	736	736
North-eastern	532-3	Well	Oak	166	570	712	736	736
North-eastern	533-3	Well	Oak	92	610	702		-722
North-eastern	801-3	Well	Oak	182	485	667	687	
		Well	Beech	189	546	735	735	735
Northern	10212-17	Well	Beech	111	650	761	762	
		Well	Beech	72	689	761	762	762
		Well	Oak	22	733	755	775	775
Northern	10290-17	Well	Beech	86	668	754	755	
		Well	Beech	52	703	755	755	
		Well	Beech	29	730	759	760	
		Well	Beech	109	651	760	761	
		Well	Beech	50	711	761	761	
		Well	Beech	65	696	761	761	
		Well	Beech	48	713	761	761	
		Well	Beech	73	688	761	761	
		Well	Beech	114	646	760	761	
		Well	Beech	90	671	761	761	
		Well	Beech	91	670	761	762	
		Well	Beech	51	710	761	762	
		Well	Beech	59	702	761	762	
		Well	Beech	124	637	761	762	
		Well	Beech	56	705	761	762	
		Well	Beech	84	677	761	762	
		Well	Beech	90	671	761	762	
		Well	Beech	111	650	761	762	
		Well	Beech	76	685	761	762	
		Well	Beech	53	708	761	762	
		Well	Beech	80	681	761	762	
		Well	Beech	64	698	762	762	
		Well	Beech	63	699	762	762	
		Well	Beech	76	685	761	762	762
Northern	10370-17	Well	Oak	154	584	738	758	
		Well	Beech	127	634	761	761	761
Central	81-3	Well	Beech	161	616	777	779	779
Central	152-3	Well	Oak	36	666	702		
Central	271b-3	Well	Oak	87	547	634		
Central	399-3	Well	Beech	171	603	774	774	
		Well	Beech	125	647	772	774	774
Central	401-3	Well	Beech	92	594	686		
Central	404-3	Well	Beech	196	558	754	775	
		Well	Ash	83	696	779	779	779
Southern	313-3	Well	Beech	132	585	717	767	
		Well	Beech	232	545	777	777	777
		Well	Oak	218	574	772	792	
		Well	Beech	97	712	809	809	809
Southern	327-3	Well	Beech	192	617	809	809	
		Well	Beech	156	655	811	811	811
Southern	528d-3	Well	Alder	186	544	730	759	759
Southern	632-3	Well	Beech	76	729	805	805	
		Well	Beech	273	533	806	806	806

Appendix 1 *(continued). The dendrochronologically dated samples from Groß Strömkendorf. The dating has been made by K.-U. Heußner, at the Deutsches Archäologisches Institut in Berlin. Altogether 95 samples from 28 features are dated (after* TUMMUSCHEIT *2003a).*

Area	Feature	Type of feature	Variety of tree	No. of growth rings	Beginning	End	Fell year	Building year
Southern	632b-3	Well	Beech	77	703	780	780	
		Well	Beech	81	699	780	780	780
Southern	648-3	Well	Beech	96	686	782	782	
		Well	Beech	184	599	783	783	783
Southern	676-3	Well	Oak	60	617	677	697	
		Well	Oak	68	651	719	739	
		Well	Oak	39	719	758	779	
		Well	Oak	54	728	781	782	782
Southern	785-3	Well	Beech	77	635	712	712	
		Well	Beech	208	521	729	764	
		Well	Beech	190	535	725	768	
		Well	Beech	151	635	786	786	
		Well	Beech	166	620	786	786	
		Well	Beech	113	674	787	787	
		Well	Beech	136	651	787	787	
		Well	Beech	125	662	786	787	
		Well	Beech	158	629	787	787	
		Well	Beech	266	521	787	787	
		Well	Beech	155	632	787	787	
		Well	Beech	97	690	787	787	
		Well	Beech	138	649	787	787	
		Well	Beech	157	630	787	787	
		Well	Beech	178	609	787	787	
		Well	Beech	135	652	787	787	
		Well	Beech	97	690	787	787	
		Well	Beech	146	641	787	787	
		Well	Beech	92	695	787	787	787
Southern	786-3	Well	Beech	237	529	766	766	
		Well	Oak	150	601	751	772	
		Well	Oak	109	644	753	773	
		Well	Ash	38	739	777	778	
		Well	Ash	45	740	785	785	
		Well	Ash	34	751	785	785	
		Well	Ash	60	725	785	786	786
Southern	804-3	Well	Beech	44	712	756	756	
Southern	811a-3	Well	Beech	106	611	717	721	
		Well	Pine	63	700	763	763	
		Well	Beech	137	649	786	786	786
Southern	7003	Well	Beech	205	571	776	776	776
Southern	7004	Well	Beech	215	562	777	777	
Southern	7016	Well	Beech	200	536	736	736	
		Well	Beech	211	527	738	749	749
Southern	7028,4	Post	Ash	27	706	733	734	?
Southern	1,1989	Well	Oak	163	566	729	729	

Appendix 1 *(end). The dendrochronologically dated samples from Groß Strömkendorf. The dating has been made by K.-U. Heußner, at the Deutsches Archäologisches Institut in Berlin. Altogether 95 samples from 28 features are dated (after* TUMMUSCHEIT *2003a).*

Σ(T x Wa) ÷ Wb = Xt

T = measured sherd thickness (mm)
Wa = sherd weight (g)
Wb = total weight (g)
Xt = average sherd thickness (mm)

Appendix 2. Formula for calculating the average sherd thickness.

Wt x 10 ÷T = Wc

Wt = true weight (g)
T = sherd thickness (mm)
10 = standard thickness (mm)
Wc = adjusted weight (g)

Appendix 3. Formula for calculating adjusted weight.

Wb ÷ (Xt x Number of sherds x 2) = S

Wb = total weight (g)
Xt = average sherd thickness (mm)
S = mean sherd size (cm^2)

Appendix 4. Formula for calculating the mean sherd size.

No.	Production place
1	Mayen, Kr. Mayen-Koblenz
2	Meckenheim, Rhein-Sien-Kreis
3	Bonn
4	Bornheim-Waldorf, Rhein-Sieg-Kreis
5	Bornheim-Sechtem, Rhein-Sieg-Kreis
6	Bornheim-Walberberg, Rhein-Sieg-Kreis
7	Brühl-Eckdorf, Erftkreis
8	Brühl-Badorf, Erftkreis
9	Krefeld-Gellep, City of Krefeld
10	Gesecke, Kr. Lippstadt, Nordrhein-Westfalen
11	Huy an der Maas
12	Brühl-Pingsdorf
13	Duisburg
14	Bad Münstereifel-Mutscheid, Krs. Euskirchen
15	Andenne an der Maas
16	Siegburg, Rhein-Sieg-Kreis
17	Siegburg, Rhein-Sieg-Kreis, Aulgasse and Galgenberg
18	Brühl-Pingsdorf, Erftkreis
19	Brühl-Kierberg, Erftkreis
20	Brühl, Erftkreis
21	Erfstadt-Liblar, Erftkreis
22	Kerpen-Horrem, Erftkreis
23	Bergisch-Gladbach-Paffrath, Rheinisch-Bergischer Kreis
24	Bergisch-Gladbach-katterbach Rheinisch-Bergischer Kreis
25	Ratingen-Hösel, Kr. Düsseldorf-Mettmann
26	Duisburg, Stadtteil Pulverberg
27	Übach-Palenberg, Kr. Heinsberg
28	Brunssum, Prov. Limburg
29	Schinveld, Prov. Limburg
30	Selfkant-Saeffelen, Kr. Heinsberg
31	Wassenberg-Effeld, Kr. Heinsberg
32	Wassenberg-Birgelen Kr. Heinsberg
33	Wassenberg-Wildenrath Kr. Heinsberg
34	Brüggen-Oebel, Kr. Viersen
35	Arnhem-Oosterbeek
36	Koblenz-Urbar, Kr. Mayen-Koblenz
37	Höhr-Grenzhausen, Höhr, Kr. Montabaur
38	Höhr-Grenzhausen, Grenzhausen Kr. Montabaur
39	Bendorf-Sayen, Kr. Mayen-Koblenz
40	Siegburg, Aulgasse, Rhein-Sieg-Kreis
41	Wachtberg-Alendorf, Rhein-Sieg-Kreis
42	Meckenheim, Wiesenweg und Uhlgasse, Rhein-Sieg-Kreis
43	Alfter-Witterschlick, Rhein-Sieg-Kreis
44	Bonn-Lengsdorf, centre of town
45	Bonn-Lengsdorf, South of Lengsdorfer Bach
46	Bonn-Poppelsdorf
47	Brühl-centre of town
48	Hürth-Fischenich, Erftkreis
49	Cologne, Maximenenstraße
50	Cologne, Komödienstraße
51	Cologne, Eigelstein
52	Frechen, Erftkreis
53	Langerwehe-Merode Kr. Düren
54	Langerwehe-centre of town Kr. Düren
55	Langerwehe-Jüngersdorf, Kr. Düren
56	Raeren
57	Neudorf
58	Kettenis
59	Eyenatten
60	Hanset
61	Aachen
62	Übach-Palenberg, Kr. Heinsberg, Teverener Heide
63	Niederkrüchten-Elempt, Kr. Viersen
64	Niederkrüchten-Overhetfeld, Kr. Viersen
65	Düsseldorf
66	Xanten, Kr. Wesel

Appendix 5. Known production sites of pottery in the area close the Rhine and Eifel (Key to Fig. 15).

Shape group	Find number	Type of feature	Area	Vessel type	Rim diameter (cm)	Base diameter (cm)	Height (cm)	Thin section	Dating	Plate No.
a	92/234/31,8	Pit and ditch		Feldberg	11	10	20			
a	97/87/238b,75	Pit		Feldberg	18	8	18	121		11,9
a	97/88/590b,10	Urn grave		Feldberg	20	8	16	61		9,1
a	98/126/10432,29	Pit		Feldberg	22	9	22			11,1
a	98/126/782a,3	Urn grave		Feldberg	13	8	14	53		9,2
a	98/126/782d,1	Urn grave		Feldberg	14	7.5	11.5	64		9,3
a	98/126/910a,29	Skeleton grave		Scandinavian	14	8	13.5			21,1
a	90/314,492	Pit		Sukow	16	9	17	141		
a	92/234/35,10	Pit house	Central	Sukow	19	9	18			
a	94/536	Grave		Sukow	18	10	16			1,1
a	95/542/122,4	Urn grave		Sukow	13	8	12	54		1,2
a	97/88/10121,24	Pit		Sukow	10	8	10	131		
a	97/88/454,1	Skeleton grave		Sukow	15	8	17.5			
a	97/88/547a,1	Pit		Sukow	21	10	24.5			4,14
a	97/88/655,4	Urn grave		Sukow	14	8	13.5			
a	97/88/735c,1	Urn grave		Sukow	15	7	15			1,3
a	97/88/738,1	Urn grave		Sukow	15	9	18.5			1,5
a	98/126/10330,265	Pit		Sukow	12	8	11			4,2
a	98/126/10375,123	Pit house	Northern	Sukow	11	8	10			4,3
a	98/126/1152,14	Cultural layer and ditch		Sukow	12	8	14			4,1
a	98/126/753c,1	Urn grave		Sukow	13.5	7	12.5			1,6
a	98/126/782e,3	Urn grave?		Sukow	11	9	14			1,7
b	96/303/476,1; 96/303/476,4	Urn grave		Feldberg	11	7	13.5			9,4
b	98/126/801b,1	Urn grave		Feldberg	14	8.5	14			9,5
b	96/302,2899	Surface find		Saxon	20	11	21	49		22,5
b	98/126/753k,1	Urn grave		Saxon	15	8	20	59		22,1
b	95/542/94,6	Urn grave		Sukow	16	9	14.5	140		1,8
b	97/88/737,1	Urn grave		Sukow	17	8	17.5			
b	97/88/742,1	Urn grave		Sukow	20	10	17.5			1,9
b	98/125/436g,7	Pit		Sukow	11	10	19.5			
b	98/126/753a,1	Urn grave		Sukow	16	10	14	60		1,10
b	98/126/803,2	Urn grave		Sukow	17	8	18	56		1,11
c	97/88/735a,1	Urn grave		Saxon	14	8	12			22,2
c	95/542/206,240	Boat- and urn grave		Sukow	13	7	15			2,1
c	95/542/79,2	Urn grave		Sukow	12.5	7	11			2,2
c	97/88/735a,2	Urn grave		Sukow	12	8	8.5	67		
d	96/303/457,1	Urn grave		Feldberg		10				9,6
d	97/88/558,1	Urn grave		Feldberg	20	13	16	68		9,7
d	97/88/693a,2	Urn grave		Feldberg	14	8	15			9,8
d	98/125/481,4; 98/125/481,13	Posthole		Feldberg	11	7	8.5			11,2
d	98/126/10310,38	Well?	Northern	Feldberg	11	7	12			
d	98/126/931b,3	Pit		Feldberg	10	7	9			11,3
d	95/542/73,4	Urn grave		Sukow	14	8	18.5			2,3
d	97/87/224,41	Pit		Sukow	23	10	22.5	97		4,5
d	97/88/556,1	Urn grave		Sukow	11	7	15			2,4
d	98/125/785,12	Well	Southern	Sukow	8	5	7	144	787	4,4
d	98/126/790,3	Urn grave		Sukow	14	9	14	52		2,5
d	98/126/854a,2	Urn grave		Sukow	7	8	14			2,6
e	96/303/330,1	Coloured surface		Feldberg	12	5	10.5			11,19
e	98/125/722,3; 98/125/722,8	Ditch		Slavonic	13	10	15			
e	70/56d			Sukow	12	7	11.5			
e	90/314,206			Sukow	18	9	16.5			

Appendix 6 *(continued). The result of the shape analyses of the 100 complete vessels from Groß Strömkendorf.*

Shape group	Find number	Type of feature	Area	Vessel type	Rim diameter (cm)	Base diameter (cm)	Height (cm)	Thin section	Dating	Plate No.
e	95/542/107,1	Cremation grave		Sukow	12	8	12.5	138		2,8
e	95/542/125,2	Skeleton grave		Sukow	11	7	12			2,9
e	95/542/88,3	Urn grave		Sukow	12	8	12	55		2,7
e	96/302/81,544	Well	Central	Sukow	8	5	6.6	142	779	
e	96/303/444,1	Urn grave		Sukow	13	7	14			2,10
e	96/519,2	Surface find		Sukow	12	9	13			
e	97/87/204,7	Pit		Sukow	9	7	11	89		4,6
e	97/88/545,2	Urn grave		Sukow	11	8	12			2,11
e	97/88/561,3	Culture layer		Sukow	13	9	12			
e	97/88/719d,1	Mound		Sukow	14	8	12	65		
e	98/125/681,10	Pit house	Southern	Sukow	12	7	13			7,4
e	98/126/1136,1; 98/126/1136,2	Pit house	Northeastern	Sukow	22	10	23			6,1
f	98/126/779b,1; 98/126/779b,2	Urn grave		Feldberg	11	7	17			9,10
f	98/126/787,3	Urn grave		Feldberg	8	7	11			9,9
f	95/542/206,234	Boat- and urn grave		Frisian	13	11	25	1		23,1
f	98/126/949,23	Mound		Ladoga?	10	8	10	145		26,3
f	96/303/270,1	Cremation grave		Sukow	6	5	6.8			2,13
f	98/126/792,1; 98/126/792,2	Urn grave		Sukow	14	8	11			
g	98/126/797,1	Urn grave		Feldberg	9	5.5	8.5			9,11
g	96/302/53,85	Pit		Sukow	16	8	14.5	87		4,7
g	96/303/451,1	Urn grave		Sukow	12	7	9.5	139		2,14
g	97/87/240,225	Pit house	Central	Sukow	10	6	10	88		4,8
g	97/87/260,102	Well	Central	Sukow	24	12	12	111		5,1
g	98/125/656,205	Well	Southern	Sukow	13	7	10.5	62		4,9
g	98/126/1134,1	Ditch		Sukow	16	8	14.5	69		4,13
g	98/126/753h,1	Urn grave		Sukow	15	9	12			2,15
h	96/303/402,1	Urn grave		Feldberg	10	5	10			9,12
h	96/303/453,1	Urn grave		Feldberg	17	9	13			9,13
h	90/314,275			Sukow	10	6	7.5	143		
h	95/541/42,2	Posthole		Sukow	14	9	14	123		4,10
h	96/303/10012,8	Pit		Sukow	12	6	11			
h	96/303/10029,9	Pit		Sukow	10	8	11			
h	96/303/418,2	Grave – Deposit		Sukow	16	9	18.5			1,4
h	96/303/465,1	Urn- and horse grave		Sukow	13	8	10.5			1,12
i	96/303/441,12	Coloured surface		Menkendorf	12	7	11			15,9
i	97/88/719,6	Mound		Menkendorf	14	10	14.5			15,10
j	98/125/436c,15	Pit		Import	10	9	6			26,1
j	96/302/59,61	Pit		Scandinavian	12	12	11	29		
j	96/303/10050,2	Pit		Scandinavian	10	9	14.2			
k	98/126/10245,39	Pit		Frisian	27	10	22	151		23,2
k	98/126/10290,102	Well	Northern	Scandinavian	10	7	10.5	152	762	
u	97/77/741,1	Urn grave		Sukow	20	11	9			3,1
u	97/87,449	Surface find		Sukow	10	8	7	158		
u	98/125/475,65	Pit		Sukow	6	5	6			4,11
u	98/126/10346c,20	Pit		Sukow	34	12	24			
u	98/126/10410,4	Ditch		Sukow	14	10	7.5			4,12
u	98/126/800,1	Urn grave		Merovingian	14	10	21	7		24,1

***Appendix 6** (end). The result of the shape analyses of the 100 complete vessels from Groß Strömkendorf.*

Thin section	Ware group	Find No.	Type of feature	Vessel type	Plate No.	Sorting	Coarse-ness	Sand	Silt	Mica	Ore	Iron oxide	Iron hy-droxide	Organic material	Calcium carbonate	Temper material	temper amount	Max. grain size (mm)	Notes
1	III	95/542/206,234	Boat-/urn grave	Frisian	23,1	S	Fine		x	++	*	*	x			Sand	10	1.5	
2	I	97/87,724	Surface find	Tating		S	Fine		x	+	-	+	x			Natural		1.0	Rich in quartz
3	I	97/87/276,120	Pit	Tating		S	Fine		x	+	*	+	x			Natural		1.3	Rich in quartz
4	VIII	98/125/242,445	Pit house	Merovingian		S	Med. coarse	x	x	+	-	+	x			Natural		1.0	Rich in quartz
5	X	97/87/161,31	Well	Badorf		S	Med. coarse		x	++	+	*	x			Sand	6	0.3	
6	X	97/88/10115,143	Ditch	Badorf	25,4	S	Med. coarse		x	++	*	*	n.o.			Sand	5	0.3	Rich in quartz
7	III	98/126/800,1	Urn grave	Merovingian	24,1	S	Fine			n.o.		-	-	x		Sand	5	0.3	
8	III	98/125/436c,13	Pit	Merovingian		S	Fine			n.o.	+	-	-	x		Sand	5	0.3	
9	IV	98/125/681,1	Pit house	Merovingian		S	Fine			n.o.	-	-	-			Sand/Grog	8	1.3	
10	VIII	98/125/436,137	Pit	Tating		S	Med. coarse	x	x	+	-	-	x	x		Natural		0.5	
11	XIII	98/125/436,476	Pit	Merovingian		S	Coarse	x		+	-	+	x			Natural		1.5	Rich in quartz
12	V	97/87/359,38	Well	Muschelgrus		S	Fine			-	*	+	x			Sand/Shells/Grog	17	0.5	
13	V	97/87/370,14	Pit house	Muschelgrus		S	Fine			-	-	+	x			Sand/Shells/Grog	18	1.0	
14	V	98/125/723,19	Pit	Muschelgrus		S	Fine			-	-	+	x			Sand/Shells/Grog	15	0.8	
15	VI	95/541/15,4	Pit	Feldberg	11,7	S	Fine		x	+	-	+	x			Granite	15	2.0	Cracked
16	VI	95/541/6,6	Hearth	Sukow	5,15	S	Fine		x	-	-	+	x			Granite/Sandstone	13	2.0	Levigated
17	VI	95/541/1,49	Pit house	Sukow	5,9	S	Fine		x	+	*	+	x			Granite	11	2.5	
18	VI	95/541/1,70	Pit house	Sukow	6,17	S	Fine		x	+	*	+	x			Granite	8	2.0	
19	VI	95/541/1,105	Pit house	Sukow	5,2	S	Fine		x	++	*	+	n.o.			Granite	17	2.5	Cracked
20	VI	95/541/1,105	Pit house	Sukow	5,10	S	Fine		x	*	-	+	x			Granite	18	2.5	Cracked
21	VII	95/541/28,41a	Pit house?	Feldberg	12,2	S	Fine		x	*	*	+	x	x		Granite/Grog	15	2.0	
22	VI	95/541/28,44a	Pit house?	Sukow	5,6	S	Fine		x	*	-	*	x			Granite	13	2.5	Cracked
23	VI	95/541/29,14	Well	Slavonic	17,1	S	Fine		x	*	-	+	x			Granite	10	2.0	
24	XI	95/541/46,4	Pit	Sukow	7,7	S	Med. coarse		x	+	-	+	x			Granite	10	2.0	
25	XIII	95/541,241	Surface find	Scandinavian		U	Coarse	x	x	*	-	+	n.o.			Natural		2.0	
26	VII	96/302/53,75	Pit	Sukow		S	Fine		x	+	+	+	x			Granite/Grog	9	2.0	
27	III	96/302/53,83	Pit	Feldberg		S	Fine			-	-	+	x			Sand	20	2.0	Cracked
28	VI	96/302/59,59	Pit	Feldberg		S	Fine		x	+	-	+	x			Granite	15	3.0	
29	XIII	96/302/59,61	Pit	Scandinavian		U	Coarse	x	x	+	*	+	x	x		Natural		1.5	
30	V	96/302/68,10	Pit	Muschelgrus		S	Fine			*	-	+	x			Sand/Shells/Grog	20	2.5	
31	VII	96/302/74,6	Pit	Sukow		S	Fine		x	+	-	+	x	x		Granite Grog	15	2.5	Cracked
32	VI	96/302/62,174	Pit	Slavonic		S	Fine		x	+	*	*	x		x	Granite	10	1.5	
33	VII	96/302/62,202	Pit	Feldberg		S	Fine		x	+	*	+	x	x		Granite/Grog	10	2.0	Cracked
34	VII	96/302/62,218	Pit	Sukow		S	Fine		x	+	*	*	x	x		Granite/Grog	16	3.0	Weathered
35	XIII	96/302/59,86	Pit	Scandinavian		U	Coarse	x	x	+	-	+	x			Natural		2.5	
36	VI	96/302/63,282	Pit	Lid (slavonic)		S	Fine		x	+	-	*	x		x	Granite	15	2.0	
37	XI	96/302/63,302	Pit	Sukow		S	Med. coarse	x	x	+	*	+	x			Granite	13	3.0	
38	VI	96/302/63,255	Pit	Menkendorf		S	Fine			+	-	+	x	x		Granite	15	2.5	Weathered
39	V	96/302/63,421	Pit	Muschelgrus		S	Fine			*	*	+	x			Sand/Shells/Grog	19	2.5	

***Appendix** 7 (continued). Results of microscopy of ceramic thin sections from Groß Strömkendorf.*
*Abbrevations: ++=very rich amount, +=rich amount, *=normal amount, -=low amount, x=presence, n.o= not observed*

Thin section	Ware group	Find No.	Type of feature	Vessel type	Plate No.	Sorting	Coarse-ness	Sand	Silt	Mica	Ore	Iron oxide	Iron hy-droxide	Organic material	Calcium carbonate	Temper material	temper amount	Max. grain size (mm)	Notes
40	VI	96/302/65,453	Pit house	Slavonic		S	Fine		x	*	*	+	x	x		Granite	17	3.5	Weathered
41	VII	96/302/103,32	Pit house	Feldberg		S	Fine		x	+	*	+	x			Granite/Grog/Sandstone	14	2.0	
42	XIII	96/302/96,119	Pit and hearth	Scandinavian		U	Coarse	x	x	+	*	+	x			Natural		3.5	
43	XIII	96/302/96,122	Pit and hearth	Sukow		U	Coarse	x	x	*	+	*	x			Natural		2.5	
44	XIV	96/302/96,124	Pit and hearth	Scandinavian		U	Coarse	x	x	-	*	n.o.	x			Sandstone	10	1.5	
45	VI	96/302/81,518	Well	Sukow		S	Fine			+	*	+	x			Sandstone	22	2.5	
46	XI	96/302/81,573	Well	Scandinavian		S	Med. coarse		x	++	*	+	x			Granite	15	2.5	Cracked
47	XIII	96/302/144,175	Pit house	Feldberg		U	Coarse	x	x	+	-	+	x			Natural		1.5	
48	XIII	96/302/144,212	Pit house	Scandinavian		U	Coarse	x	x	*	-	+	x			Natural		2.0	
49	VII	96/302,2899	Surface find	Saxon	22,5	S	Fine		x	+	+	+	x			Granite/Grog/Sandstone	13	3.5	
50	XII	96/302,2899	Surface find	Sukow	5,3	S	Med. coarse		x	++	*	*	x			Granite/Grog	22	3.0	Weathered
51	VII	96/302/142,7	Pit	Sukow		S	Fine		x	+	-	*	x	x		Granite/Grog	11	2.0	
52	VI	98/126/790,3	Urn grave	Sukow	2,5	S	Fine		x	+	n.o.	+	x			Granite	17	2.5	
53	VII	98/126/782a,3	Urn grave	Feldberg	9,2	S	Fine		x	-	-	++	x			Granite/Grog/(Flint)	11	2.5	
54	VI	95/542/122,4	Urn grave	Sukow	1,2	S	Fine		x	++	+	++	x			Granite/Sandstone	13	2.0	
55	VI	95/542/88,3	Urn grave	Sukow	2,7	S	Fine		x	+	-	+	x			Granite/Sandstone	22	2.5	
56	VI	98/126/803,2	Urn grave	Sukow	1,11	S	Fine		x	++	-	+	x			Granite	11	3.0	
57	VI	98/126/769,232	Boat grave	Scandinavian	21,4	S	Fine		x	*	-	*	x			Granite/Sandstone	18	2.5	Weathered
58	XI	98/126/802,1	Urn grave	Slavonic	20,1	S	Med. coarse		x	++	-	+	x			Granite/Flint	8	3.0	Cracked
59	VII	98/126/753k,1	Urn grave	Saxon	22,1	S	Fine		x	++	*	+	x	x		Granite/Grog	9	2.5	
60	VI	98/126/753a,1	Urn grave	Sukow	1,10	S	Fine		x	*	*	+	x			Granite	25	4.0	Weathered
61	VI	97/88/590b,10	Urn grave	Feldberg	9,1	S	Fine		x	*	-	+	x	x		Granite	12	2.5	
62	VII	98/125/656,205	Well	Sukow	4,9	S	Fine		x	+	-	*	x			Granite/Grog	10	1.5	Cracked
63	VII	98/126/803,4	Urn grave	Feldberg	10,1	S	Fine		x	*	-	+	x			Granite/Grog	15	4.0	
64	VI	98/126/782d,1	Urn grave	Feldberg	9,3	S	Fine			+	*	+	x			Granite	10	1.5	
65	VI	97/88/719d,1	Mound	Sukow		S	Fine		x	+	*	+	x			Granite	14	2.5	
66	VI	97/88/631,3	Urn grave	Jutland?	21,5	S	Fine			-		+	x			Granite (gneis)	17	3.0	Levigated
67	VI	97/88/735a,2	Urn grave	Sukow		S	Fine		x	+		*	x			Granite	17	3.0	Weathered
68	XIII	97/88/558,1	Urn grave	Feldberg	9,7	S	Coarse		x			+	x			Natural		3.0	
69	VI	98/126/1134,1	Ditch	Sukow	4,13	S	Fine			*	-	+	x			Granite	8	1.5	
70	VI	98/126/931i,1	Pit	Feldberg	10,6	S	Fine		x	++	-	*	n.o.			Granite	20	4.5	
71	XIII	97/87/212,96	Pit	Sukow	8,9	S	Coarse	x	x	+	*	+	x	x		Natural		2.0	
72	VI	97/87/212,96	Pit	Feldberg	12,3	S	Fine		x	+	-	+	x			Granite	10	3.0	
73	VI	97/87/253,88	Pit house	Sukow		S	Fine		x	*	-	*	x		x	Granit	12	2.5	
74	VI	95/542/114,1	Skeleton grave	Scandinavian		S	Fine		x	*	-	*	x			Granite	5	3.5	Weathered
75	VIII	96/302/167,239	Pit house	Crucible		S	Med. coarse		x	+	*	+	x			Natural		1.0	
76	VI	96/302/167,394	Pit house	Feldberg		S	Fine		x	+	-	+	x	x		Granite	13	2.5	Levigated
77	XI	96/302/153,23	Pit house	Slavonic		S	Med. coarse		x	++	n.o.	+	x			Granite	21	3.5	Weathered
78	XIII	96/302/163a,2	Pit	Scandinavian		U	Coarse	x	x	-	n.o.	+	x		x	Natural		2.5	Jutland?

***Appendix** 7 (continued). Results of microscopy of ceramic thin sections from Groß Strömkendorf.*
*Abbrevations: ++=very rich amount, +=rich amount, *=normal amount, -=low amount, x=presence, n.o= not observed*

Thin section	Ware group	Find No.	Type of feature	Vessel type	Plate No.	Sorting	Coarse-ness	Sand	Silt	Mica	Ore	Iron oxide	Iron hy-droxide	Organic material	Calcium carbonate	Temper material	temper amount	Max. grain size (mm)	Notes
79	XII	96/302/163a,1	Pit	Slavonic		S	Med. coarse	x	x	++	*	+	x			Granite/Grog	9	2.5	
80	VI	96/302/164,3	Pit and hearth	Import?		S	Fine		x	+	n.o.	*	x			Granite	6	1.5	
81	XIII	97/87/238b,7	Pit	Scandinavian		U	Coarse	x	x	*	-	*	x			Natural		2.5	
82	XI	97/87/238c,4	Pit	Local?		S	Med. coarse		x	++	-	*	x	x	x	Granite	9	1.5	
83	VI	97/87/404,10	Well	Sukow		S	Fine		x	++	-	+	x			Granite	10	3.5	
84	VI	95/542/133,1	Pit	Feldberg	12,1	S	Fine		x	*	-	+	x			Granite	8	3.0	
85	VI	96/302/166,113	Pit	Menkendorf		S	Fine		x	+	-	+	x		x	Granite	13	2.5	Fossile
86	X	96/302/166,127	Pit	Menkendorf		S	Med. coarse		x	+	n.o.	+	x			Sand	12	2.5	Fossile
87	VI	96/302/53,85	Pit	Sukow	4,7	S	Fine		x	*	n.o.	+	x			Granite	15	2.5	
88	XI	97/87/240,225	Pit house	Sukow	4,8	S	Med. coarse		x	+	*	+	x			Granite	11	2.0	
89	III	97/87/204,7	Pit	Sukow	4,6	S	Fine		x	++		+	x			Sand	15	2.5	Weathered
90	VI	97/87/240,224	Pit house	Fresendorf	15,7	S	Fine		x	+	*	+	x			Granite	10	1.5	
91	VI	97/87/240,232	Pit house	Sukow		S	Fine		x	++	-	*	x			Granite	6	2.0	Levigated
92	VI	97/87/215,220	Pit house	Import?		S	Fine		x	++	-	+	x			Granite	7	1.5	
93	VI	97/87/224,23	Pit	Slavonic	17,12	S	Fine		x	+	-	+	x			Granite	12	3.0	Levigated
94	VI	97/87/224,44	Pit	Fresendorf	15,5	S	Fine		x	++	-	+	x			Granite	7	2.5	Weathered
95	XIII	97/87/224,42	Pit	Sukow	5,7	U	Coarse	x	x	++	-	*	x			Natural		2.0	
96	VI	97/87/224,44	Pit	Feldberg		S	Fine		x	++	*	+	x			Granite	20	2.5	
97	VI	97/87/224,41	Pit	Sukow	4,5	S	Fine		x	++	*	+	x	x		Granite	15	3.0	
98	VI	97/87,925	Surface find	Slavonic	17,2	S	Fine		x	++	-	*	x			Granite	17	3.0	
99	VI	97/87/313,4	Well	Feldberg	13,4	S	Fine		x	+	*	+	x			Granite	10	2.5	
100	VI	97/87/313,33	Well	Menkendorf	16,5	S	Fine		x	+	*	+	x			Granite/Sandstone	26	3.0	Weathered
101	VI	97/87/313,33	Well	Fresendorf	15,6	S	Fine		x	*	*	+	x			Granite	10	3.0	
102	VI	97/87/313,143	Well	Sukow		S	Fine		x	+	-	+	x			Granite	10	2.0	
103	XIII	97/87/313,144	Well	Sukow		U	Coarse	x	x	+	-	+	x			Natural		1.0	
104	VI	97/87/313,159	Well	Feldberg	13,14	S	Fine		x	++	-	+	x	x		Granite	12	3.5	
105	VI	97/87/313,161	Well	Scandinavian		S	Fine			-	-	+	x			Granite	17	3.5	
106	VI	97/87/313,243	Well	Feldberg	13,5	S	Fine		x	+	-	+	x			Granite	23	2.5	Levigated
107	VI	97/87/313,324	Well	Sukow		S	Fine		x	*	-	+	x			Granite	16	2.0	Levigated
108	XIII	97/87/313,324	Well	Scandinavian		U	Coarse	x	x	*	*	+	x			Natural		1.5	
109	VI	97/87/359,35	Well	Feldberg	14,9	S	Fine		x	-	*	+	x			Sandstone	18	3.5	Calcium carbonate
110	XIII	97/87/366,10	Pit house	Sukow		U	Coarse	x	x	+	*	+	x			Natural		1.5	
111	VI	97/87/260,102	Well	Sukow	5,1	S	Fine		x	+	*	+	x			Granite/Sandstone	15	2.5	Cracked Levigated
112	XIII	97/87/263,14	Culture layer	Scandinavian		U	Coarse	x	x	-	*	+	x			Natural		1.5	
113	VII	97/87/206,134	Pit house	Scandinavian		S	Fine			++	-	+	x			Granite/Grog	11	2.0	
114	VII	97/87/206,119	Pit house	Feldberg	14,14	S	Fine		x	+	-	+	x			Sandstone/Grog	12	2.5	Cracked
115	XII	97/87/206,256	Pit house	Sukow	5,17	S	Med. coarse		x	++	*	+	x			Granite/Grog	12	2.5	Cracked
116	XII	97/87/206,387	Pit house	Fresendorf	15,2	S	Med. coarse		x	++	-	+	n.o.			Granite/Grog	11	2.5	
117	VI	97/87/206,261	Pit house	Feldberg	14,3	S	Fine		x	+	-	+	x	x		Granite	20	2.5	

***Appendix** 7 (continued). Results of microscopy of ceramic thin sections from Groß Strömkendorf.*
*Abbrevations: ++=very rich amount, +=rich amount, *=normal amount, -=low amount, x=presence, n.o= not observed*

Thin section	Ware group	Find No.	Type of feature	Vessel type	Plate No.	Sorting	Coarse-ness	Sand	Silt	Mica	Ore	Iron oxide	Iron hy-droxide	Organic material	Calcium carbonate	Temper material	temper amount	Max. grain size (mm)	Notes
118	VII	97/87/206,261	Pit house	Sukow	5,18	S	Fine		x	++	*	+	x			Sandstone/Grog	8	2.5	Cracked
119	XIII	97/87/206,136	Pit house	Scandinavian		U	Coarse	x	x	+	-	+	x			Natural		1.5	
120	XIII	97/87/238b,75	Pit	Frisian	23,3	U	Coarse	x	x	-	-	+	x			Natural		2.5	
121	XIII	97/87/238b,75	Pit	Feldberg	11,9	U	Coarse	x	x	*	-	+	x		x	Natural		2.5	
122	XIII	97/87/242,29	Pit house	Scandinavian		U	Coarse	x	x	+	-	+	x			Natural		2.5	
123	VI	95/541/42,2	Posthole	Sukow	4,10	S	Fine		x	+	-	+	x			Granite	16	3.0	
124	XI	96/303/241,1	Mound	Iron Age	26,5	S	Med. coarse		x	++	-	*	x			Granite	12	3.0	
125	XIII	96/303/465,12	Urn-/horse grave	Scandinavian	21,2	U	Coarse	x	x	*	-	+	x			Natural		1.5	
126	XV	96/303/311,4	Pit	Ember vessel		S	Coarse	x	x	++	*	*	x	x		Organic		0.5	
127	VI	96/303/284,6	Cremation grave	Menkendorf	16,3	S	Fine		x	++	*	+	x			Granite	13	3.5	
128	VIII	97/87/252,115	Irregular pit		26,9	S	Med. coarse	x	x	++	-	+	x			Natural		1.5	Granofyrian texture
129	VI	97/87/206,387	Pit house	Jutland		S	Fine		x	++	*	+	x	x		Granite	11	3.0	
130	VI	97/87/206,387	Pit house	Menkendorf	16,8	S	Fine		x	++	-	+	x			Granite/Sandstone	17	3.0	
131	XIV	97/88/10121,24	Pit	Sukow		S	Coarse	x	x	++		+	x		x	Granite	10	2.5	
132	XI	97/88/489,37	Pit house	Slavonic	26,8	S	Med. coarse		x	++	-	n.o.	n.o.			Granite	6	2.0	
133	VI	97/88/498,26	Pit	Slavonic	26,7	S	Fine		x	++	-	+	x			Granite	15	2.5	
134	V	98/126/10332a,4	Pit	Local?		S	Fine		x	++		+	x			Sand/Shells/Grog	10	1.5	
135	VI	98/126/10370,89	Well	Scandinavian		S	Fine		x	++		+	x			Granite	23	2.5	Cracked
136	XIII	98/126/10388,1	Pit	Import?		S	Coarse	x	x	++	*	+	x		x	Natural		2.5	
137	VI	98/126/10385,205	Pit house	Import	26,6	S	Fine		x	++	-	+	x			Granite	13	2.5	
138	VI	95/542/107,1	Cremation grave	Sukow	2,8	S	Fine		x	++	-	+	x			Granite/Sandstone	10	2.5	
139	XIII	96/303/451,1	Urn grave	Sukow	2,14	U	Coarse	x	x	+	*	+	x			Natural		2.5	
140	XIII	95/542/94,6	Urn grave	Sukow	1,8	U	Coarse	x	x	++	*	+	x			Natural		2.0	
141	VI	90/314,492	Pit	Sukow		S	Fine		x	+		+	x			Granite	20	4.5	
142	VII	96/302/81,544	Well	Sukow		S	Fine		x	++		+	x			Granite/Grog	9	1.5	
143	XIII	90/314,275		Sukow		U	Coarse	x	x	++	+	+	x			Natural		2.0	
144	VI	98/125/785,12	Well	Sukow	4,4	S	Fine		x	++	*	+	x		x	Granite	10	2.5	
145	XI	98/126/949,23	Mound	Ladoga?	26,3	S	Med. coarse	x	x	++		+	x			Granite	17	4.5	
146	VI	98/126/10372,5	Pit	Slavonic	26,4	S	Fine		x	++		*	x	x		Granite	8	2.5	
147	VII	98/126/10201a,137	Well	Hanging pot	21,10	S	Fine			-	-	*	x			Granite/Grog/Sandstone	11	2.0	
148	VI	98/126/10370,66	Well	Scandinavian		S	Fine		x	++	-	+	x			Granite	17	5.0	Levigated
149	V	98/126/10308,7	Pit	Muschelgrus		S	Fine		x	*		+	x			Sand/Shells/Grog	11	2.5	
150	V	96/302,225	Surface find	Muschelgrus		S	Fine		x	*		+	x			Sand/Shells/Grog	11	2.0	
151	XIII	98/126/10245,39	Pit	Frisian	23,2	S	Coarse	x	x	*	-	+	x		x	Natural		1.5	
152	VI	98/126/10290,102	Well	Scandinavian		S	Fine		x	++	*	+	x			Granite	7	2.0	
153	III	98/126/10310,43	Well?	Frisian		S	Fine			-	*	+	x			Sand	10	1.0	
154	VI	98/126/1034,1	Pit	Sukow		S	Fine		x	+		+	x			Granite	13	2.5	
155	XIII	98/126/841,3	Pit	Curonian	26,2	U	Coarse	x	x	++	-	+	x			Natural		1.5	
156	XIV	96/303/10022,4	Pit house	Uncertain		U	Coarse	x	x	++	-	+	x	x		Granite	14	3.0	

***Appendix** 7 (continued). Results of microscopy of ceramic thin sections from Groß Strömkendorf.*
*Abbrevations: ++=very rich amount, +=rich amount, *=normal amount, -=low amount, x=presence, n.o= not observed*

Thin section	Ware group	Find No.	Type of feature	Vessel type	Plate No.	Sorting	Coarse-ness	Sand	Silt	Mica	Ore	Iron oxide	Iron hy-droxide	Organic material	Calcium carbonate	Temper material	temper amount	Max. grain size (mm)	Notes
157	IX	97/88/10115,130	Ditch	Merovingian		S	Med. coarse	x	x	*	-	*	x	x		Natural/Grog		1.5	
158	XIII	97/87,449	Surface find	Sukow		U	Coarse	x	x	++		+	x			Natural		2.0	
159	XIII	96/518,4	Surface find	Feldberg		U	Coarse	x	x	++		+	x			Natural		1.5	
160	VIII	98/125/632,235	Well	Badorf		S	Med. coarse	x	x	-	*	*	x			Natural		0.3	
161	I	98/125,61	Surface find	Merovingian		S	Fine		x		+	*	x			Natural		0.5	
162	IV	97/87/215,307	Pit house	Mayen		S	Fine		x	++	+	*	x			Sand/Grog	10	1.0	Volcanic
163	I	96/518,6	Surface find	Mayen		S	Fine		x		+	*	x			Natural		1.0	Volcanic
164	II	98/126/10357,96	Pit	Walberberg		S	Fine		x	++	-	*	x			Natural/Grog		1.0	Rich in quartz
165	IX	98/126/791,1	Urn grave?	Mayen		S	Med. coarse		x	*	+	*	x			Natural/Grog		1.5	Volcanic
166	IV	98/126/10398,248	Pit house	Badorf		S	Fine		x	*	-	*	x			Sand/Grog	5	1.0	Rich in quartz
167	XIII	97/87/256,118	Pit	Tating		S	Coarse	x	x	-	+	+	x			Natural		1.0	
168	VIII	98/125,156	Surface find	Tating		S	Med. coarse		x	-	-	*	x			Natural		1.3	
169	VIII	96/303/10022,32	Pit	Merovingian		S	Med. coarse		x	*	*	*	x			Natural		0.3	
170	VIII	98/126/10440b,6	Pit	Badorf		S	Med. coarse	x	x	++	+	*	x			Natural		0.5	
171	IX	98/125/507,17	Well	Mayen		S	Med. coarse		x	-	+	*	n.o.			Natural/Grog		0.5	
172	IV	97/87/256,130	Pit-concentration	Merovingian		S	Fine			-	-	*	x			Sand/Grog	6	1.5	Volcanic
173	XIII	90/314,426 ,427		Merovingian		S	Coarse	x	x	-	-	*	x			Natural		1.0	
174	XIII	98/126/1061,12	Pit - Iron Age?	Frankish		U	Coarse	x	x	+	*	+	x	x		Natural		1.5	
175	VIII	98/125/498,153	Pit	Badorf		S	Med. coarse	x	x	-	-	*	x			Natural		0.5	
176	III	92/234/45,7	Well	Frisian	23,6	S	Fine			+		+	x			Quartzsand	18	1.0	
177	III	97/87/246,25	Well?	Frisian	23,5	S	Fine			+		+	x			Quartzsand	24	1.8	

***Appendix** 7 (end). Results of microscopy of ceramic thin sections from Groß Strömkendorf.*
*Abbrevations: ++=very rich amount, +=rich amount, *=normal amount, -=low amount, x=presence, n.o= not observed*

Plate No.	Find No.	Type of feature	Area	Vessel type	Thin section	Dating	Shape group
1,1	94/536	Grave		Sukow			a
1,2	95/542/122,4	Urn grave		Sukow	54		a
1,3	97/88/735c,1	Urn grave		Sukow			a
1,4	96/303/418,2	Grave – Deposit		Sukow			h
1,5	97/88/738,1	Urn grave		Sukow			a
1,6	98/126/753c,1	Urn grave		Sukow			a
1,7	98/126/782e,3	Urn grave?		Sukow			a
1,8	95/542/94,6	Urn grave		Sukow	140		b
1,9	97/88/742,1	Urn grave		Sukow			b
1,10	98/126/753a,1	Urn grave		Sukow	60		b
1,11	98/126/803,2	Urn grave		Sukow	56		b
1,12	96/303/465,1	Urn- and horse grave		Sukow			h
2,1	95/542/206,240	Boat- and urn grave		Sukow			c
2,2	95/542/79,2	Urn grave		Sukow			c
2,3	95/542/73,4	Urn grave		Sukow			d
2,4	97/88/556,1	Urn grave		Sukow			d
2,5	98/126/790,3	Urn grave		Sukow	52		d
2,6	98/126/854a,2	Urn grave		Sukow			d
2,7	95/542/88,3	Urn grave		Sukow	55		e
2,8	95/542/107,1	Cremation grave		Sukow	138		e
2,9	95/542/125,2	Skeleton grave		Sukow			e
2,10	96/303/444,1	Urn grave		Sukow			e
2,11	97/88/545,2	Urn grave		Sukow			e
2,12	97/88/678a,1	Urn grave		Sukow			e
2,13	96/303/270,1	Cremation grave		Sukow			f
2,14	96/303/451,1	Urn grave		Sukow	139		g
2,15	98/126/753h,1	Urn grave		Sukow			g
3,1	97/88/741,1	Urn grave		Sukow			u
3,2	97/88/518,24	Urn grave		Sukow			
3,3	95/542/62,4	Grave		Sukow			
3,4	98/126/830,20	Mound		Sukow			
3,5	96/303/282,2 ; 96/303/282,5 ; 96/303/282,9	Grave – Deposit		Sukow			
3,6	97/88/545,1	Urn grave		Sukow			
3,7	95/542/122,3			Sukow			
3,8	98/126/792,1	Urn grave		Sukow			
3,9	98/126/753,27	Mound		Sukow			
3,10	98/126/802,8			Sukow			
3,11	98/126/802,8			Sukow			
3,12	95/542/132,14			Sukow			
3,13	98/126/802,8	Urn grave		Sukow			
3,14	95/542/90,1	Cremation place		Sukow			
3,15	97/88/658,11	Cultural layer and mound		Sukow			
4,1	98/126/1152,14	Cultural layer and ditch		Sukow			a
4,2	98/126/10330,265 267	Pit		Sukow			a
4,3	98/126/10375,123	Pit house	Northern	Sukow			a
4,4	98/125/785,12	Well	Southern	Sukow	144	787	d
4,5	97/87/224,41	Pit		Sukow	97		d
4,6	97/87/204,7	Pit		Sukow	89		e
4,7	96/302/53,85	Pit		Sukow	87		g
4,8	97/87/240,225	Pit house	Central	Sukow	88		g
4,9	98/125/656,205	Well	Southern	Sukow	62		g
4,10	95/541/42,2	Posthole		Sukow	123		h
4,11	98/125/475,65	Pit		Sukow			u
4,12	98/126/10410,4	Ditch		Sukow			u
4,13	98/126/1134,1	Ditch		Sukow	69		g
4,14	97/88/547a,1	Pit		Sukow			a
5,1	97/87/260,102	Well	Central	Sukow	111		g
5,2	95/541/1,105	Pit house	Central	Sukow	19		
5,3	96/302,2899	Surface find		Sukow	50		
5,4	97/87/256f,35	Pit		Sukow			
5,5	97/88/553,18 ; 97/88/553,25	Cultural layer and pit house	Northeastern	Sukow			
5,6	95/541/28,44a	Pit house?		Sukow	22		

***Appendix 8** (continued). List of the pottery found at Groß Strömkendorf as illustrated on the plates 1–26.*

Plate No.	Find No.	Type of feature	Area	Vessel type	Thin section	Dating	Shape group
5,7	97/87/224,42	Pit		Sukow	95		
5,8	98/125/436,133	Pit		Sukow			
5,9	95/541/1,49	Pit house	Central	Sukow	17		
5,10	95/541/1,105	Pit house	Central	Sukow	20		
5,11	98/126/10370,81	Well	Northern	Sukow		761	
5,12	98/126/1035,15	Cultural layer and ditch		Sukow			
5,13	97/88/10156a,62 ; 97/88/10156a,14	Cooking pit		Sukow			
5,14	98/126/1052,6	Pit house	Northeastern	Sukow			
5,15	95/541/6,6	Hearth		Sukow	16		
5,16	97/88/739,14	Pit		Sukow			
5,17	97/87/206,256	Pit house	Central	Sukow	115		
5,18	97/87/206,261	Pit house	Central	Sukow	118		
6,1	98/126/1136,1 ; 98/126/1136,2	Pit house	Northeastern	Sukow			e
6,2	97/88/10156a,61 ; 97/88/10156a,14	Cooking pit		Sukow			
6,3	97/88/10130a,100	Hearth / cooking pit		Sukow			
6,4	98/126/10368,65	Pit house	Northern	Sukow			
6,5	98/126/10398,4	Pit house		Sukow			
6,6	97/88/10114,65	Pit		Sukow			
6,7	98/126/10311,31	Pit		Sukow			
6,8	98/125/784,16	Well	Southern	Sukow			
6,9	98/125/633,56	Well	Southern	Sukow			
6,10	98/125/638,375 ; 98/125/638,378	Well and pit house	Southern	Sukow			
6,11	97/88/10126,3	Pit house?		Sukow			
6,12	95/542/200,1	Pit house	Northeastern	Sukow			
6,13	98/126/10375b,105 ; 98/126/10375b,35	Pit house	Northern	Sukow			
6,14	98/125/718,382	Ditch		Sukow			
6,15	97/87/240,220	Pit house	Central	Sukow			
6,16	98/126/765,6	Pit		Sukow			
6,17	95/541/1,70	Pit house	Central	Sukow	18		
6,18	97/87/402,8	Pit		Sukow			
7,1	96/303/436g,2	Coloured surface		Sukow			
7,2	98/126/10360,169	Pit		Sukow			
7,3	98/126/10375j,23	Pit		Sukow			
7,4	98/125/681,10	Pit house	Southern	Sukow			e
7,5	97/88/554,12	Pit		Sukow			
7,6	98/126/822,6	Pit		Sukow			
7,7	95/541/46,4	Pit		Sukow	24		
7,8	92/234/31,9	Pit		Sukow			
7,9	97/88/739,29	Pit		Sukow			
7,10	98/126/1052,22	Pit house	Northeastern	Sukow			
7,11	98/126/10440,117	Cultural layer with pit and ditch		Sukow			
7,12	95/541/30,28	Pit house	Central	Sukow			
7,13	95/541,308	Surface find		Sukow			
8,1	96/303/346,6	Pit		Sukow			
8,2	97/87/224,23	Pit		Sukow			
8,3	98/125/718,356	Ditch		Sukow			
8,4	98/125/436,501	Pit		Sukow			
8,5	98/126/846,16	Pit		Sukow			
8,6	97/87/498,5	Pit		Sukow			
8,7	92/234/31,9	Pit		Sukow			
8,8	96/303/346,5	Pit		Sukow			
8,9	97/87/212,96	Pit		Sukow	71		
8,10	98/126/10218,12 ; 98/126/10218,7	Pit house	Northern	Sukow			
8,11	92/234/31,9	Pit		Sukow			
9,1	97/88/590b,10	Urn grave		Feldberg	61		a
9,2	98/126/782a,3	Urn grave		Feldberg	53		a
9,3	98/126/782d,1	Urn grave		Feldberg	64		a
9,4	96/303/476,1 ; 96/303/476,4	Urn grave		Feldberg			b
9,5	98/126/801b,1	Urn grave		Feldberg			b
9,6	96/303/457,1	Urn grave		Feldberg			d
9,7	97/88/558,1	Urn grave		Feldberg	68		d
9,10	98/126/779b,1 ; 98/126/779b,2	Urn grave		Feldberg			f
9,11	98/126/797,1	Urn grave		Feldberg			g
9,12	96/303/402,1	Urn grave		Feldberg			h

***Appendix 8** (continued). List of the pottery found at Groß Strömkendorf as illustrated on the plates 1–26.*

Bibliography

Alexander 1973
C. Alexander, Notes on the Synthesis of Form (Cambridge 1973).

Ambrosiani / Clarke 1998
B. Ambrosiani / H. Clarke, Birka and the beginning of the Viking Age. In: A. Wesse (Hrsg.), Studien zur Archäologie des Ostseeraumes. Von der Eisenzeit zum Mittelalter. Festschrift für Michael Müller-Wille (Neumünster 1998) 33–38.

Andersen 1975
S. H. Andersen, Ringkloster, en jysk inlandsboplats med Ertebøllekultur. Kuml 1973–1974, 1975, 11–108.

Arbman 1940
H. Arbman, Birka I. Die Gräber. Tafeln (Stockholm 1940).

Arnold 1985
D. E. Arnold, Ceramic theory and cultural process. New Stud. Arch. (Cambridge 1985).

Arnold 1993
D. E. Arnold, Ecology and ceramic production in an Andean Community. New Stud. Arch. (Cambridge 1993).

Arwidsson 1984
G. Arwidsson, Glas. In: G. Arwidsson (Hrsg.), Birka II. 1. Systematische Analysen der Gräberfunde (Stockholm 1984) 203–212.

Bäck 1995
M. Bäck, Importkeramiken i Birka. Medeltitsark. tidskr. 95,1, 1995, 4–21.

Bäck in prep.
M. Bäck, Pottery from the Baltic and Beyond. Birka Studies. Eastern Connections 3 (in preparation).

Bardet 1995
A. C. Bardet, Pottery Traded to Dorestad: some Exploratory Archaeometrical Analyses of Early Medieval Rhenish Wares. Ber. ROB 41, 1995, 187–251.

Barford 2001
P. M. Barford, The Early Slavs (London 2001).

Bender Jørgensen 1992
L. Bender Jørgensen, North European Textiles until AD 1000 (Aarhus 1992).

Bergmann 1993
R. Bergmann, Mittelalterliche und neuzeitliche Töpferei in Dortmund-Groppenbruch. Vorbericht. Ausgr. u. Funde Westfalen-Lippe. 8 B, 1993, 31–40.

Birkhoff 1933
G. D. Birkhoff, Aesthetic measure (Cambridge 1933).

Bøe 1931
J. Bøe, Jernalderens Keramikk i Norge. Bergens Mus. Skr. 14 (Bergen 1931).

Böhner 1955/56
K. Böhner, Frühmittelalterliche Töpferöfen in Walberberg und Pingsdorf. Bonner Jahrb. 155/156, 1955/1956, 372–387.

Brather 1993
S. Brather, Altslawische Keramik in Mecklenburg und Vorpommern. Probleme der Typenverbreitung. Zeitschr. Arch. 27,1, 1993, 323–335.

Brather 1996
S. Brather, Feldberger Keramik und frühe Slawen. Studien zur nordwestslawischen Keramik der Karolingerzeit. Univforsch. Prähist. Arch. 34 (Berlin 1996).

Brather 2001
S. Brather, Archäologie der westlichen Slawen. Siedlung, Wirtschaft und Gesellschaft im früh- und hochmittelalterlichen Ostmitteleuropa. Ergbd. RGA 30 (Berlin, New York 2001).

Brather 2004
S. Brather, Ethnische Interpretationen in der frühgeschichtlichen Archäologie. Geschichte, Grundlagen und Alternativen. Ergbd. RGA 42 (Berlin 2004).

Broberg / Hasselmo 1981
B. Broberg / M. Hasselmo, Keramik, kammar och skor från 7 medeltida städer. Fyndstudie. Medeltidsstaden 30 (Stockholm 1981).

Brorsson 1999
T. Brorsson, Keramik. In: B. Jacobsson (ed.), Trelleborgen i Trelleborg. Förhistoriska boplatslämningar och gravar, vikingatida ringborg och medeltida bebyggelselämningar. Skåne, Trelleborg, kv Gröningen, kv Kråkvinkeln, Bryggaregatan och Svenstorpsgatan. UV Syd Rapport 1999:93. Lund.

Brorsson 2000
T. Brorsson, Keramik från yngre järnålder och tidig medeltid. In: F. Svanberg / B. Söderberg (eds.), Porten till Skåne. Löddeköpinge under järnålder och medeltid. Riksantikvarieämbetet, Ark. undersökningar, Skr. 32 (Lund 2000) 188–225.

Brorsson 2001
T. Brorsson, Behind the Pottery: Signs of Contacts across the Baltic Sea. In: M. Brisbane / D. Gaimster (eds.), Novgorod. The archaeology of a Russian Medieval City and its Hinterland. British Mus. Occasional Paper 141 (London 2001) 51–58.

Brorsson 2002a
T. Brorsson, Den vendel- och vikingatida keramiken. In: C. B. Pettersson, "Bott vid en landsväg". In: M. Mogren (ed.), Märkvärt, medeltida. Arkeologi ur en lång skånsk historia. Riksantikvarieämbetet, Ark. undersökningar, Skr. 43 (Lund 2002) 74–81.

BRORSSON 2002b
T. BRORSSON, Ceramics as an identity marker – a preliminary study on the Groß Strömkendorf pottery. Bodendenkmalpfl. Mecklenburg-Vorpommern Jahrb. 50, 2002, 247–256.
BRORSSON 2003a
T. BRORSSON, Analys av Keramik från den vikingatida handelplatsen Kaupang, Larvik, Vestfold. KFL Rapport 03/1219. Report at the Laboratory for Ceramic Research, Lund University. Lund.
BRORSSON 2003b
T. BRORSSON, Keramik från en centralplats. In: Söderberg 2003, 341–372.
BRORSSON 2003c
T. BRORSSON, Pottery from Early Viking age graves in the Baltic Region. Towards the Interpretation of a Society. Bodendenkmalpfl. Mecklenburg-Vorpommern Jahrb. 51, 2003, 361–374.
BRORSSON 2003d
T. BRORSSON, The Slavonic Feldberg and Fresendorf pottery in Scania, Sweden. In: B. Hårdh / L. Larsson (eds.), Centrality – Regionality. The social structure of Southern Sweden during the iron age. Acta Arch. Lundensia in 8°,40 (Lund 2003) 223–234.
BRORSSON 2005
T. BRORSSON, Byns keramik. In: M. Mogren (ed.), Byarnas bönder. Medeltida samhällsförändring i Västskåne. Skånska spår, arkeologi längs Västkustbanan. Riksantikvarieämbetet, Ark. undersökningar, Skr. 30 (Lund 2005) 294–321.
BRORSSON 2006
T. BRORSSON, Keramiken. In: S. Larsson, Gårdar på rad – Centraliteter. Människor, Strategier och Landskap. Riksantikvarieämbetet 1:1, 62–67. Lund.
BRORSSON / STANISŁAWSKI 1999
T. BRORSSON / B. STANISŁAWSKI, Ceramika słowiańska typu Feldberg i Fresendorf w Skanii, na tle produkcji garncarskiej wczesnośredniowiecznego Wolina. Mat. Zachidniopomorskie XLV, 1999, 283–312.
BUKO / PELA 1997
A. BUKO / W. PELA (EDS.), Imported and Locally Produced Pottery: Methods of Identification and Analysis. Materials from I Conference of Pottery Research Group of Committee for Pre- and Protohistoric Sciences of Polish Academy of Sciences (Warszawa 1997).
CALLMER 1988
J. CALLMER, Slawisch-skandinavische Kontakte am Beispiel der slawischen Keramik in Skandinavien während des 8. und 9. Jahrhunderts. Ber. RGK 69, 1988, 654–674.
CALLMER 1991
J. CALLMER, Platser med anknytning till handel och hantverk i yngre järnålder. In: P. Mortensen / B. Rasmussen (eds.), Fra Stamme til Stat i Danmark 2. Høvdingesamfund og Kongemagt. Jysk Ark. Selskabs Skr. 22,2 (Aarhus 1991) 29–47.
CALLMER 1992
J. CALLMER, Interaction between Ethnical Groups in the Baltic Region in the Late Iron Age. In: B. Hårdh / B. Wyszomirska-Werbart (eds.), Contacts Across the Baltic Sea during the Late Iron Age (5th–12th centuries). Baltic Sea Conference, Lund, October 25–27, 1991. Univ. Lund, Inst. Arch., Report Ser. 43 (Lund 1992) 99–108.
CAPELLE 1998
T. CAPELLE, Die Sachsen des frühen Mittelalters (Stuttgart 1998).
CARLSSON 1982
K. CARLSSON, Importkeramik i Gamla Lödöse. Lödöse – Västsvensk medeltidsstad 3,2 (Stockholm 1982).
CLARKE / AMBROSIANI 1993
H. CLARKE / B. AMBROSIANI, Vikingastäder (Höganäs 1993).
DAIM 2000
F. DAIM, "Byzantinische" Gürtelgarnituren des 8. Jahrhunderts. In: F. Daim (Hrsg.), Die Awaren am Rand der byzantinischen Welt. Studien zu Diplomatie, Handel und Technologietransfer im Frühmittelalter. Monogr. Frühgesch. u. Mittelalterarch. 7 (Innsbruck 2000) 77–204.
DULINICZ 2006
M. DULINICZ, Frühe Slawen im Gebiet zwischen unterer Weichsel und Elbe. Eine archäologische Studie. Stud. Siedlungsgesch. Arch. Ostseegebiete 7 (Neumünster 2006).
EISENSCHMIDT 2004
S. EISENSCHMIDT, Grabfunde des 8. bis 11. Jahrhunderts zwischen Kongeå und Eider. Zur Bestattungssitte der Wikingerzeit im südlichen Altdänemark. Stud. Siedlungsgesch. u. Arch. Ostseegebiete 5,1–2 (Neumünster 2004).
ELLMERS 1972
D. ELLMERS, Frühmittelalterliche Handelsschiffahrt in Mittel- und Nordeuropa. Offa 28 (Neumünster 1972).
ERDMANN ET AL. 2001
W. ERDMANN / H. J. KÜHN / H. LÜDTKE / E. RING / W. WESSEL, Rahmenterminologie zur Mittelalterlichen Keramik in Norddeutschland. In: Lüdtke / Schietzel 2001, 947–969.
VAN ES / VERWERS 1980
W. A. VAN ES / W. J. H. VERWERS, Excavations at Dorestad 1. The Harbour: Hoogstraat I. Nederlandse Oudheden 9 (Amersfoort 1980).
FEVEILE / JENSEN 2000
C. FEVEILE / S. JENSEN, Ribe in the 8th and 9th century. A Contribution to the Archaeological Chronology of North Western Europe. Acta Arch. (København) 71=Suppl. 2 (2000) 9–24.
FEVEILE ET AL. 1998
C. FEVEILE / S. JENSEN / K. LUND RASMUSSEN, Produktion af drejat keramik i Ribeområdet i sen yngre germansk jernalder. Proveniensbestemmelse ved hjælp af magnetisk susceptibilitet og termoluminiscens. Kuml 1997–1998, 1998, 143–159.
FILIPOWIAK / GUNDLACH 1992
W. FILIPOWIAK / H. GUNDLACH, Wolin Vineta. Die tatsächliche Legende vom Untergang und Aufstieg der Stadt (Rostock 1992).
VON FRIESEN 1976
O. VON FRIESEN, Krukan från Raus (Uddevalla 1976).
GABRIEL 1984
I. GABRIEL, Starigard/Oldenburg. Hauptburg der Slawen in Wagrien (Neumünster 1984).
Geologische Karte von Mecklenburg-Vorpommern
Landesamt für Umwelt, Naturschutz und Geologie (Hrsg.), Übersichtskarte 1 : 500 000, Oberfläche (Güstrow 2000).
GERDS 2006
M. GERDS, Scandinavian burial rites on the southern Baltic coast. In: A. Andrén / K. Jennbert / C. Raudvere (eds.), Old Norse religion in long-term perspectives. Vägar till Midgård 8 (Lund 2006) 153–158.
GERDS in prep.
M. GERDS, The Cemetery at Groß Strömkendorf (in preparation).

Gojda 1991
M. Gojda, The Ancient Slavs. Settlement and Society. The rhind lectures 1989–90 (Edinburgh 1991).
Gosselain / Livingstone Smith 1995
O. Gosselain / A. Livingstone Smith, The ceramics and society project. An ethnographic and experimental approach to technological choices. In: A. Lindahl / O. Stilborg (eds.), The Aim of Laboratory Analyses of Ceramics in Archaeology. April 7–9, 1995 in Lund, Sweden. In Honour of Birgitta Hulthén ass. prof. emer. Konferenser 34 (Stockholm 1995) 147–160.
Gräslund 1980
A.-S. Gräslund, Birka IV. The Burial Customs. A Study of the graves on Björkö (Stockholm 1980).
Gustafsson 1995
S. Gustafsson, Vad har kärlen använts till? – analyser av 4 krukskärvor av keramik från bosättningen vid Birkas stadsvall (unpublished essay, Archaeological Research Laboratory, Stockholm University 1995).
Hahn 1977
K.-D. Hahn, Die einheimische Keramik von Haithabu (Diss. Kiel 1977).
Helgesson 1996
B. Helgesson, Fjälkinge 35:60 m fl. Fjälkinge socken. Fornlämning 18 och 19. Rapport Arkeologisk undersökning 1990. Kristianstads Läns Museum (Kristianstad 1996).
Herrmann 1970
J. Herrmann, Die Slawen in Deutschland. Ein Handbuch (Berlin 1970).
Hodges 1981
R. Hodges, The Hamwih pottery: the local and imported wares from 30 years' excavations at Middle Saxon Southampton and their European context. Council British Arch. Research Report 37 (London 1981).
Holand 2003
I. Holand, Pottery. In: G. Stamsø Munch / O. S. Jansen / E. Roesdahl (eds.), Borg in Lofoten. A chieftain's farm in North Norway. Ark. Skr. 1 (Trondheim 2003) 199–212. Trondheim.
Hougen 1969
E. K. Hougen, Lerkarmaterialet fra Kaupang. Viking 33, 1969, 97–118.
Hougen 1993
E. K. Hougen, Kaupang-funnene, bd. IIIb. Bosetningsområdets keramikk (Oslo 1993).
Hulthén 1974
B. Hulthén, On Documentation of Pottery. Acta Arch. Lundensia Ser. in 8°, minore 3 (Lund 1974).
Hulthén 1976
B. Hulthén, On Thermal Colour Test. Norwegian Arch. Rev. 9,1, 1976, 1–6.
Hulthén 1977
B. Hulthén, On Ceramic Technology During the Scanian Neolithic and Bronze Age. Theses and Papers North-European Arch. 6 (Stockholm1977).
Hulthén 1978
B. Hulthén, Keramiktillverkning i Kv Tankbåten i Ystad. Ystadiana. Ystads Fornm. 23, 1978, 103–113.
Hulthén 1984
B. Hulthén, Keramik. In: G. Arwidsson (Hrsg.), Birka II. 1. Systematische Analysen der Gräberfunde (Stockholm 1984) 249–262.
Hulthén 1985
B. Hulthén, Temper variations in ancient ceramics. – Technological or cultural origin? In: T. Edgren (ed.), Proceedings of Third Nordic Conference on the Application of Scientific Methods in Archaeology, Mariehamn, Åland, Finland 8–11 October 1984. ISKOS 5, 1985, 329–337.
Hulthén 1995
B. Hulthén, Ceramics and Clays at Ancient Högom. A ceramological study. Ant. Arkiv 77 (Stockholm 1995).
Isaksson 2000
S. Isaksson, The culture of food in Early Medieval Middle Sweden. A pottery use perspective. In: S. Isaksson, Food and Rank in Early Medieval Time. Theses and Papers Scien. Arch. 3 (Stockholm 2000).
Jagodziński 2006
M. F. Jagodziński, RGA^2 31 s.v. Truso (Berlin/New York 2006) 291–295.
Jankuhn 1937
H. Jankuhn, Die Wehranlagen der Wikingerzeit zwischen Schlei und Treene (Neumünster 1937).
Janssen 1970
W. Janssen, Der karolingische Töpferbezirk von Brühl-Eckdorf, Kreis Köln. Neue Ausgr. u. Forsch. Niedersachsen 6 (Hildesheim 1970) 224–239.
Janssen 1987
W. Janssen, Die Importkeramik von Haithabu (Neumünster 1987).
Jantzen / Schirren 1998
D. Jantzen / M. Schirren, „Rerik steht wieder auf" oder „Die Lösung des Reric-Problems" im April 1938. In: A. Wesse (Hrsg.), Studien zur Archäologie des Ostseeraumes. Von der Eisenzeit zum Mittelalter. Festschrift für Michael Müller-Wille (Neumünster 1998) 67–76.
Jöns 1998a
H. Jöns, Der frühgeschichtliche Seehandelsplatz von Groß Strömkendorf. In: C. Lübke (Hrsg.), Struktur und Wandel im Früh- und Hochmittelalter. Eine Bestandsaufnahme aktueller Forschungen zur Germania Slavica (Stuttgart 1998) 127–143.
Jöns 1998b
H. Jöns, Groß Strömkendorf-handelsplatsen ved Wismar – et exempel på urbaniseringens tidlige fase i det vestslaviske område. In: D. Meier (ed.), Beretning fra syttende tværfaglige vikingesymposium (Højbjerg 1998) 31–54.
Jöns 2000
H. Jöns, Handel und Handwerk im westslawischen Siedlungsgebiet am Beispiel des Handelsplatzes von Groß Strömkendorf bei Wismar. In: F. Lüth / U. Schonknecht (Hrsg.), Schutz des Kulturerbes unter Wasser. Veränderungen europäischer Lebenskultur durch Fluß- und Seehandel. Beiträge zum Internationalen Kongreß für Unterwasserarchäologie (IKUWA '99), 18.–21. Februar 1999 in Sassnitz auf Rügen. Beitr. Ur- u. Frühgesch. Mecklenburg-Vorpommern 35 (Lübstorf 2000) 399–403.
Jöns 2006
H. Jöns, Slawisch-skandinavische Kontakte an der südwestlichen Ostseeküste im Lichte neuer Forschungen. In: M. Bertašius (ed.), Transformatio mundi. The transition from the Late Migration period to the Early Viking Age in the East Baltic (Kaunas 2006) 117–132.

Jørgensen 1990
L. Jørgensen, Bækkegård and Glasergård. Two Cemeteries from the Late Iron Age on Bornholm. Ark. Stud. 3 (København 1990).

Keller 1998
C. Keller, Karolingerzeitliche Töpferöfen in Bornheim-Walberberg, Rhein-Sieg-Kreis. Bonner Jahrb. 198, 1998, 285–348.

Keller 2004a
C. Keller, Badorf, Walberberg und Hunneschans. Zur zeitlichen Gliederung karolingerzeitlicher Keramik vom Köln-Bonner Vorgebirge. Arch. Korrbl. 34,1, 2004, 125–137.

Keller 2004b
C. Keller, Pottery production in Badorf and Walberberg during the Carolingian period. In: R. Simek / U. Engel (eds.), Vikings on the Rhine. Recent Research on Early Medieval Relations between the Rhinelands and Scandinavia. Studia Mediaevalia Septentrionalia 11 (Wien 2004) 155–164.

Kelm 2000
R. Kelm, Mölleholmen. Eine slawische Inselsiedlung des 11. Jahrhunderts in Schonen, Südschweden. Univ. Lund Inst. Arch. Report Ser. 74 (Lund 2000).

Kempke 1981a
T. Kempke, Frühmittelalterliche Keramik aus Oldenberg in Holstein. Text (Diss. Hamburg 1981).

Kempke 1981b
T. Kempke, Frühmittelalterliche Keramik aus Oldenberg in Holstein. Tabellen, Diagramme, Karten, Tafeln (Diss. Hamburg 1981).

Kempke 1984
T. Kempke, Starigard/Oldenburg. Hauptburg der Slawen in Wagrien. 2. Die Keramik des 8.–12. Jahrhunderts. Offa 53 (Neumünster 1984).

Kempke 2002
T. Kempke, Slawische Keramik vom Hamburger Domplatz. In: R. Busch / O. Harck (Hrsg.), Domplatzgrabung in Hamburg 2. Veröff. Helms-Museum, Hamburger Mus. Arch. u. Gesch. Harburg 89 (Neumünster 2002) 95–152.

Kleingärtner / Jöns 2000
S. Kleingärtner / H. Jöns, Eine Pferdefibel aus Groß Strömkendorf, Lkr. Nordwestmecklenburg. Bodendenkmalpfl. Mecklenburg-Vorpommern Jahrb. 48, 2000, 279–293.

Kluge-Pinsker 1985
A. Kluge-Pinsker, The Pottery Supply of Duisburg, Rhineland, in the Ninth-Tenth century. Medieval Ceramics 9, 1985, 51–56.

Knorr 1970
H. A. Knorr, Handwerk und Gewerbe. In: Herrmann 1970, 73–103.

Krüger 1967
B. Krüger, Dessau-Mosigkau. Ein frühslawischer Siedlungsplatz im mittleren Elbegebiet. Schr. Sektion Vor- u. Frühgesch. 22 (Berlin 1967).

Laing 2003
L. Laing, Pottery in Britain 4000 BC to AD 1900. A Guide to identifying Pot Sherds (Witham 2003).

Laux 1997
F. Laux, Studien zur frühgeschichtlichen Keramik aus dem slawischen Burgwall bei Hollenstadt, Ldkr. Harburg. Hammaburg N. F. 11, 1997, 7–184.

Laux 2003
F. Laux, Das Steingrab in der Fischbeker Heide. Hammaburg N. F. 14, 2003, 7–178.

Lindahl 1986
A. Lindahl, Information through Sherds. A case study of the early glazed earthenware from Dalby, Scania. Lund Stud. Medieval Arch. 3 (Lund 1986).

Lindahl 2001
A. Lindahl, Laborativa analyser av Keramik från Hellum, Barmer och Kragelund. Hikuin 28, 2001, 281–296.

Lindahl 2002a
A. Lindahl, Microscopical analyses of ceramic thin-sections of votive terracottas from Tessennano, Vulci, Tuscania and Tarquinia. Appendix 1. In: M. Söderlind, Late Etruscan votive heads from Tessennano. Studia Archaeologica 118 (Rome 2002) 393–423.

Lindahl 2002b
A. Lindahl, Kärlets form och bruk. In: A. Lindahl / A. Carlie / D. Olausson (eds.), Keramik I Sydsverige. En handbok för arkeologer (Lund 2002) 38–39.

Lindahl / Matenga 1995
A. Lindahl / E. Matenga, Present and Past. Ceramics and Homesteads. An ethnoarchaeological project in the Buhera district, Zimbabwe. Stud. African Arch. 11 (Uppsala 1995).

Lobbedey 1968
U. Lobbedey, Untersuchungen mittelalterlicher Keramik vornehmlich aus Südwestdeutschland. Arbeiten Frühmittelalterforsch. 3 (Berlin 1968).

Lüdtke / Schietzel 2001
H. Lüdtke / K. Schietzel, Handbuch zur mittelalterlichen Keramik in Nordeuropa. Schr. Arch. Landesmuseum 6,1–3 (Neumünster 2001).

Łosiński / Rogosz 1984
W. Łosiński / R. Rogosz, Die Periodisierung der Frühmittelalterlichen Keramik aus Szczecin. Przegląd Arch. 32, 1984, 187–207.

Madsen 1991
H. J. Madsen, Vikingetidens keramik som historisk kilde. In: P. Mortensen / B. Rasmussen (eds.), Fra Stamme til Stat i Danmark 2. Høvdingesamfund og Kongemagt. Jysk Ark. Selskabs Skr. 22,2 (Aarhus 1991) 217–234.

Marney 1989
P. T. Marney, Roman and Belgic Pottery from excavations in Milton Keyenes 1972–82. Buckinghamshire Arch. Soc. Monogr. Ser. 2 (Aylesbury 1989).

McCarthy / Brooks 1988
M. R. McCarthy / C. M. Brooks, Medieval Pottery in Britain AD 900-1600 (Leicester 1988).

Meier 1994
Dietrich Meier, Die wikingerzeitliche Siedlung von Kosel (Kosel-West), Kreis Rendsburg-Eckernförde (Neumünster 1994).

Meier 1990
Dirk Meier, Scharstorf. Eine slawische Burg in Ostholstein und ihr Umland. Archäologische Funde. Offa-Bücher 70 (Neumünster 1990).

Müller 2002
U. Müller, Tatinger Ware im Gebiet des Mare Balticum. In: G. Mangelsdorf (Hrsg.), Aus der Frühgeschichte des südwestlichen Ostseegebietes. Greifswalder Mitt. 5 (Frankfurt a. M. 2002) 9–24.

Müller-Wille 1985
M. Müller-Wille, Westeuropäischer Import der Wikingerzeit in Nordeuropa. Society and trade in the Baltic during the Viking Age. Acta Visbyensia 7 (Visby 1985) 79–102.

Müller-Wille 1994
M. Müller-Wille, Death and Burial in Medieval Europe. Scripta Minora (Lund) 1992–1993 (Stockholm 1994).

Müller-Wille 2001
M. Müller-Wille, Ribe – Reric – Hedeby. Early Urbanization in Southern Scandinavia and the Western Slavonic Area. Offa 58, 2001, 21–31.

Müller-Wille 2002
M. Müller-Wille, Zwischen Kieler Förde und Wismarbucht. Archäologie der Obodriten vom späten 7. bis zur Mitte des 12. Jahrhunderts. Ber. RGK 83, 2002, 243–264

Müller-Wille / Tummuscheit 2004
M. Müller-Wille / A. Tummuscheit, Viking-age proto-urban centres and their hinterlands: some examples from the Baltic area. In: J. Hines / A. Lane / M. Redknap (eds.), Land, sea and home. Proceedings of a conference on Viking period settlement. Soc. Medieval Arch., Monogr. Ser. 20 (Leeds 2004) 27–39.

Munsell 1971
Munsell Soil Color Charts (Baltimore 1971).

Nerman 1958
B. Nerman, Grobin-Seeburg. Ausgrabungen und Funde (Stockholm 1958).

Nerman 1969
B. Nerman, Die Vendelzeit Gotlands 2. Tafeln (Stockholm 1969).

Nielsen 1998
L. C. Nielsen, Okholm, Vester Vedsted sogn. In: S. Jensen (ed.), Marsk, land og bebyggelse. Ribeegnen gennem 10.000 år 2. Katalog. Jysk Ark. Selskabs Skr. 35,2 (Højbjerg 1998) 173–184.

Nordström 1972
H.-Å. Nordström, Neolithic and A-Group Sites. The Scandinavian Joint Expedition to Sudanese Nubia 3 (Uppsala 1972).

Nosov 1992
E. Nosov, Ryurik Gorodishche and the settlements to the North of Lake Ilmen. In: M. A. Brisbane (ed.), The Archaeology of Novgorod, Russia. Soc. Medieval Arch., Monogr. Ser. 13 (Lincoln 1992) 5–66.

Nørlund 1948
P. Nørlund, Trelleborg. Nordisk Fortidsminder 4,1 (København 1948).

Odencrants 1933
R. Odencrants, Båtgraven nr 4 vid Valsgärde. Uppland Fornm. tidskr. 43, 1933, 225–240.

Ohlsson 1975/76
T. Ohlsson, The Löddeköpinge Investigation I. The Settlement at Vikhögsvägen. Meddel. Lund Univ. Hist. Mus. 1975–1976, 1975/76, 59–161.

Orton et al. 1993
C. Orton / P. Tyers / A. Vince, Pottery in Archaeology (Cambridge 1993).

Oxenstierna 1998
G. Oxenstierna, Nestorskrönikan. Translation of the Chronicle by Nestor (Stehag 1998).

Paddenberg 2000
D. Paddenberg, Studien zu frühslawischen Bestattungssitten in Nordostdeutschland. Offa 57, 2000, 231–345.

Pöche 2005
A. Pöche, Perlen, Trichtergläser, Tesserae. Spuren des Glashandels und Glashandwerks auf dem frühgeschichtlichen Handelsplatz von Groß Strömkendorf, Landkreis Nordwestmecklenburg. Beitr. Ur- u. Frühgesch. Mecklenburg-Vorpommern 44 (Lübstorf 2005).

Redknap 1988
M. Redknap, Medieval pottery production at Mayen: Recent advances, current problem. In: D. R. M. Gaimster / M. Redknap / H. H. Wegner (Hrsg.), Zur Keramik des Mittelalters und der beginnenden Neuzeit im Rheinland. Medieval and later pottery from the Rhineland and its markets. BAR Internat. Ser. 440 (Oxford 1988) 3–38.

Redknap 1998
M. Redknap, Late Merovingian Black and Red Burnished wares from Mayen (Rheinland-Pfalz). Arch. Korrbl. 14, 1998, 403–416.

Redknap 1999
M. Redknap, Die römischen und mittelalterlichen Töpfereien in Mayen, Kreis Mayen-Koblenz Berichte zur Arch. am Mittelrhein 6, 1999 (2003) 11–402.

Reisborg 1981a
S. Reisborg, The pottery from Building Group 3. In: A. Lundström / H. Clarke (eds.), Excavations at Helgö VII. Glass – Iron – Clay (Stockholm 1981) 135–174.

Reisborg 1981b
S. Reisborg, The pottery from Building Group 4. In: A. Lundström / H. Clarke (eds.), Excavations at Helgö VII. Glass – Iron – Clay (Stockholm 1981) 107–134.

Ring / Wieczorek 1979
E. Ring / A. Wieczorek, Tatinger Kannen aus Mainz. Arch. Korrbl. 9, 1979, 355–362.

Roslund 2001
M. Roslund, Gäster i huset. Kulturell överföring mellan slaver och skandinaver 900 till 1300 (Lund 2001).

Rydh 1936
H. Rydh, Förhistoriska undersökningar på Adelsö (Stockholm 1936).

Salskov Roberts 2002
H. Salskov Roberts, Pots for the Living, Pots for the dead. Were Pots Purpose-made for the Funeral or Reused? Can inscriptions throw light on the problem? In: A. Rathje / M. Nielsen / B. Bundgaard Rasmussen, Pots for the Living, pots for the Dead. Acta Hyperborea 9 (Copenhagen 2002) 9–32.

Samuelsson 1998
B.-Å. Samuelsson, Ljungbackagravfältet – studier av ett sydvästskånskt gravmaterial från sen järnålder (unpublished essay, Archaeological Institution, Lund University 1998).

Sanke 1999
M. Sanke, Älteste Pingsdorfer Ware aus einem Töpferbetrieb in Brühl-Pingsdorf. Ein Beitrag zur Kenntnis rheinischer Keramik des 10. Jahrhunderts. Bonner Jahrb. 199, 1999, 181–234.

Sawyer / Sawyer 2002
B. Sawyer / P. Sawyer, Die Welt der Wikinger. Die Deutschen und das Europäische Mittelalter (Berlin 2002).

Sawyer 1971
P. H. Sawyer, The Age of the Vikings (London 1971).

SCHMITZ 1999
R. SCHMITZ, Das Umfeld von Groß Strömkendorf, Ldkr. Nordwestmecklenburg, in frühslawischer Zeit und die Grabung am Burgwall Ilow, Ldkr. Nordwestmecklenburg, Fdpl. 2 (Magisterarbeit Kiel 1999).

SCHMÖLCKE 2004
U. SCHMÖLCKE, Nutztierhaltung, Jagd und Fischfang. Zur Nahrungsmittelwirtschaft des frühgeschichtlichen Handelsplatzes von Groß Strömkendorf, Landkreis Nordwestmecklenburg. Beitr. Ur- u. Frühgesch. Mecklenburg-Vorpommerns 43 (Lübstorf 2004).

SCHOKNECHT 1977
U. SCHOKNECHT, Menzlin. Ein frühgeschichtlicher Handelsplatz an der Peene (Berlin 1977).

SCHULDT 1953
E. SCHULDT, Die Slawische Keramik vom Fresendorfer Typ. Jahrb. Bodendenkmalpfl. Mecklenburg 1953 (1955) 133–150.

SCHULDT 1955
E. SCHULDT, Eine slawische Siedlung von Groß Strömkendorf, Kreis Wismar. Jahrb. Bodendenkmalpfl. Mecklenburg 1955 (1957) 135–141.

SCHULDT 1956
E. SCHULDT, Die slawische Keramik in Mecklenburg. Schr. Sektion Vor- u. Frühgesch. 5 (Berlin 1956).

SCHULDT 1963
E. SCHULDT, Die slawische Keramik von Sukow und das Problem der Feldberger Gruppe. Jahrb. Bodendenkmalpfl. Mecklenburg 1963 (1964) 239–262.

SCHULDT 1964
E. SCHULDT, Slawische Töpferei in Mecklenburg (Schwerin 1964).

SCHULDT 1980
E. SCHULDT, Handwerk und Gewerbe des 8. bis 12. Jahrhunderts in Mecklenburg (Schwerin 1980).

SCHULDT 1981
E. SCHULDT, Groß Raden. Die Keramik einer slawischen Siedlung des 9./10 Jh. (Schwerin 1981).

SELLING 1955
D. SELLING, Wikingerzeitliche und Frühmittelalterliche Keramik in Schweden (Stockholm 1955).

SELLING 1971
D. SELLING, Vikingatid – en krukmakeriets vågdal. Göreborgs arkeologiska museums årstryck 1971 (Göteborg 1971) 2–12.

SELLING 1972
D. SELLING, Review: E. K. Hougen, Leikarmaterialet fra Kaupang (The Pottery from Kaupang). Viking 33, 1969. 97–118. Norwegian Arch. Rev. 5,2, 1972, 40–44.

SHEPARD 1963
A. SHEPARD, Ceramics for the Archaeologist (Washington D.C. 1963).

SKOVGAARD-PETERSEN 1981
I. SKOVGAARD-PETERSEN, The Written Sources. In: M. Bencard (ed.), Ribe Excavations 1970–76 (Esbjerg 1981) 21–62.

SÖDERBERG 2003
B. SÖDERBERG (ed.), Järrestad. Huvudgård i centralbygd. Riksantikvarieämbetet, Ark. undersökningar, Skr. 51 (Lund 2003).

SÖDERBERG 2005
B. SÖDERBERG. Aristokratiskt rum och gränsöverskridande. Riksantikvarieämbetet, Ark. undersökningar, Skr. 62 (Lund 2005).

STEUER 1974
H. STEUER, Die Südsiedlung von Haithabu. Studien zur frühmittelalterlichen Keramik im Nordseeküstenbereich und in Schleswig-Holstein. Ausgr. Haithabu 6 (Neumünster 1974).

STEUER 1979
H. STEUER, Die Keramik aus der frühgeschichtlichen Wurt Elisenhof. In: A. Lang (Hrsg.), Die frühgeschichtliche Marschensiedlung beim Elisenhof in Eiderstedt 3 (Frankfurt a. M. 1979) 3–147.

STILBORG 1997
O. STILBORG, Shards of Iron Age Communications. A ceramological study of internal structure and external contacts in the Gudme-Lundeborg area, Funen, during the Late Roman Iron Age (Lund 1997).

STILBORG 2002
O. STILBORG, Vem var keramikern? In: A. Lindahl / A. Carlie / D. Olausson (eds.), Keramik I Sydsverige. En handbok för arkeologer (Lund 2002) 36–37.

STILBORG 2003
O. STILBORG, Late Iron Age Metal Craft Ceramics at Uppåkra. In: B. Hårdh / L. Larsson (eds.), Fler fynd i centrum (Lund 2003) 137–164.

STILKE 1995a
H. STILKE, Die früh- bis spätmittelalterliche Keramik von Emden. Probleme Küstenforsch. Südl. Nordseegebiet 22 (Oldenburg 1995) 9–202.

STILKE 1995b
H. STILKE, Shell-tempered ware from the Frisian coastal area. Medieval Ceramics 19, 1995, 11–18.

STILKE 1998
H. STILKE, Westliche Importkeramik des frühen Mittelalters im Ostseegebiet. In: A. Wesse (Hrsg.), Studien zur Archäologie des Ostseeraumes. Von der Eisenzeit zum Mittelalter. Festschrift für Michael Müller-Wille (Neumünster 1998) 577–584.

STILKE 2001
H. STILKE, Katalog der Münzschatzgefäße. In: Lüdtke / Schietzel 2001, 763–946.

STILKE ET AL. 1999
H. STILKE / A. HEIN / H. MOMMSEN, Neutronenaktivierungsanalysen an mittelalterlicher Keramik aus Mayen und an Tatinger Ware. Berichte zur Arch. am Mittelrhein 6, 1999 (2003) 404–418.

STILKE ET AL. 2001
H. STILKE / A. HEIN / H. MOMMSEN, Results of Neutron Activation Analysis on Tating ware and the Mayen Industry. Medieval Ceramics 20, 2001, 25–32.

STRÖMBERG 1978
M. STRÖMBERG, En kustby i Ystad – fore stadens tillkomst. Ystadiana. Ystads Fornm. 23, 1978, 7–102.

SUNDIUS 1955
N. SUNDIUS, Exkurs I: Mikroskopische Untersuchungen von Keramik-Scherben aus der Wikingerzeit und dem Frühen Mittelalter. In: D. Selling, Wikingerzeitliche und frühmittelalterliche Keramik in Schweden (Stockholm 1955) 244–249.

SVANBERG 2003
F. SVANBERG, Decolonizing the Viking Age 1. Acta Arch. Lundensia Ser. in 8°,43 (Lund 2003).

Tiemeyer 1995
J. Tiemeyer, Die Keramik des frühen und hohen Mittelalters aus Dalem (Landkreis Kr. Cuxhaven) und Niens (Landkreis Wesermarsch). Probleme Küstenforsch. Südl. Nordseegebiet 22 (Oldenburg 1995) 237–381.

Tummuscheit 2003a
A. Tummuscheit, Die Baubefunde des frühmittelalterlichen Seehandelsplatzes von Groß Strömkendorf, Kr. Nordwestmecklenburg (Diss. Kiel 2003).

Tummuscheit 2003b
A. Tummuscheit, Groß Strömkendorf: a Market Site of the Eighth Century on the Baltic Sea Coast. In: T. Pestell / K. Ulmschneider (eds.), Markets in Early Medieval Europe. Trading and "Productive" Sites, 650–850 (London 2003) 208–220.

Unverzagt / Schuldt 1963
W. Unverzagt / E. Schuldt, Teterow. Ein slawischer Burgwall in Mecklenburg. Schr. Sektion Vor- u. Frühgesch. 13 (Berlin 1963).

Warnke 1992
D. Warnke, Rostock-Dierkow – Ein Wirtschaftszentrum des 8./9. Jahrhunderts an der Unterwarnow. Zeitschr. Arch. Mittelalter 20, 1992, 63-80.

Watt 1983
M. A. Watt, A Viking Age Settlement at Runegård (Grødby), Bornholm. An Enterim Report of the Investigations 1979–82. Journal Danish Arch. 2, 1983, 137–148.

Wegner 1988
H. H. Wegner, Archäologische Beobachtungen zur Mittelalterlichen Keramikproduktion in Mayen, Kreis Mayen-Koblenz. In: D. R. M. Gaimster / M. Redknap / H. H. Wegner. (Hrsg.), Zur Keramik des Mittelalters und der beginnenden Neuzeit im Rheinland. Medieval and later pottery from the Rhineland and its markets. BAR Internat. Ser. 440 (Oxford 1988) 39–52.

Wietrzichowski 1989
F. Wietrzichowski, Zur Verbreitung und Entwicklung der Sukower Gruppe in Mecklenburg. Jahrb. Bodendenkmalpfl. Mecklenburg 37, 1989, 37–95.

Wietrzichowski 1993
F. Wietrzichowski, Untersuchungen zu den Anfängen des frühmittelalterlichen Seehandels im südlichen Ostseeraum unter besonderer Berücksichtigung der Grabungsergebnisse von Groß Strömkendorf. Wismarer Stud. Arch. u. Gesch. 3 (Wismar 1993).

Wilde 1939
K. A. Wilde, Die Bedeutung der Grabung Wollin 1934. Methodische Grundlagen für die Erforschung der Wikinger- und Slawen Siedlung Wollin (Stettin 1939).

Winkelmann 1984
W. Winkelmann, Liturgisches Gefäß der Missionszeit aus Paderborg. Zur Verbreitung und Deutung der Tatinger Kannen. In: W. Winkelmann (Hrsg.), Beiträge zur Frühgeschichte Westfalens. Veröff. Altkomm. Provinzialinst. Westfälische Landes- und Volkskde. 8 (Münster 1984) 129–134.

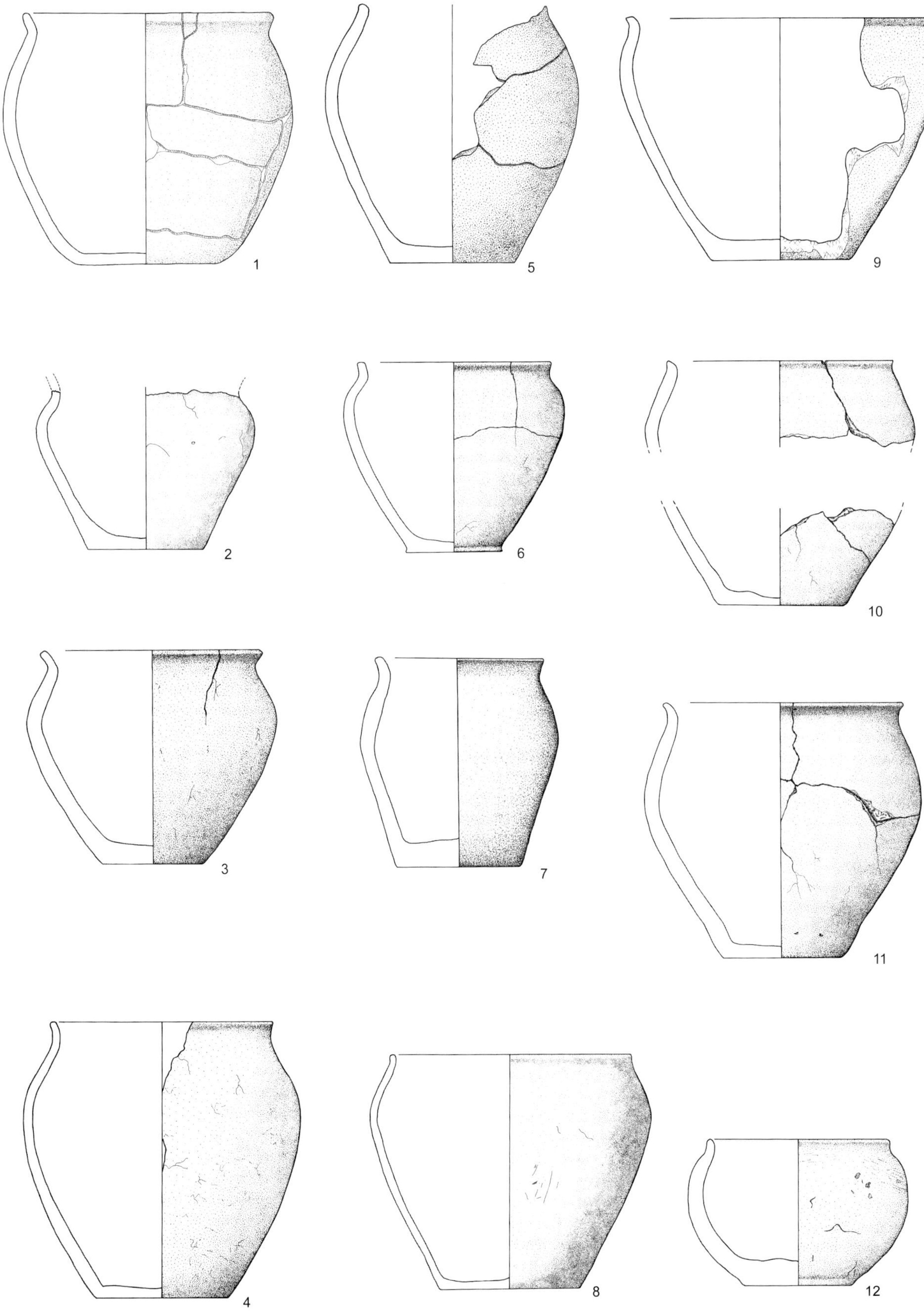

Plate 1. *Groß Strömkendorf, Lkr. Nordwestmecklenburg. Sukow pottery from the cemetery. Scale 1:4.*

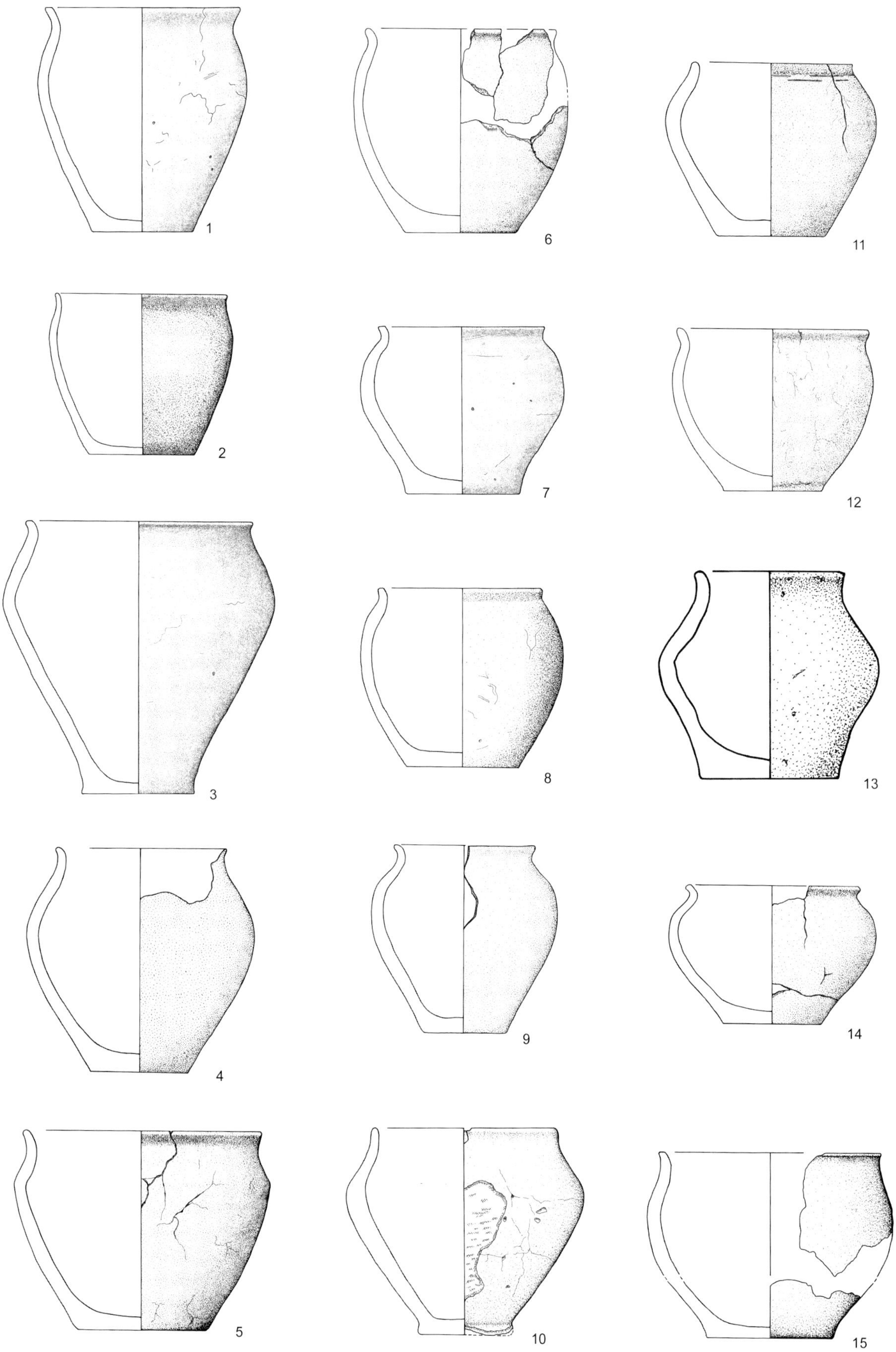

Plate 2. *Groß Strömkendorf, Lkr. Nordwestmecklenburg. Sukow pottery from the cemetery. 1–12, 14–15 scale 1:4; 13 scale 1:2.*

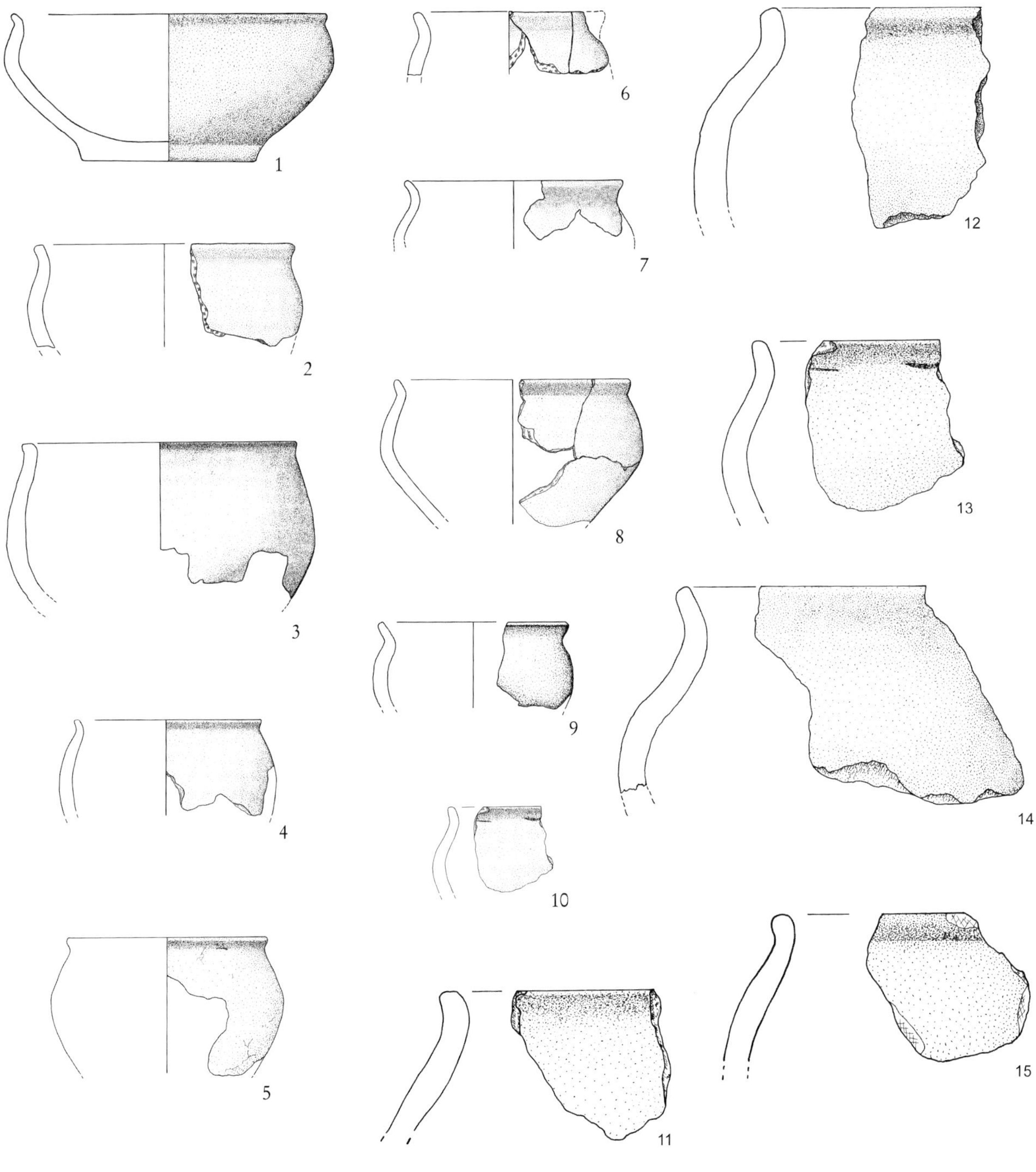

Plate 3. Groß Strömkendorf, Lkr. Nordwestmecklenburg. Sukow pottery from the cemetery. 1–9 scale 1:4; 10–15 scale 1:2.

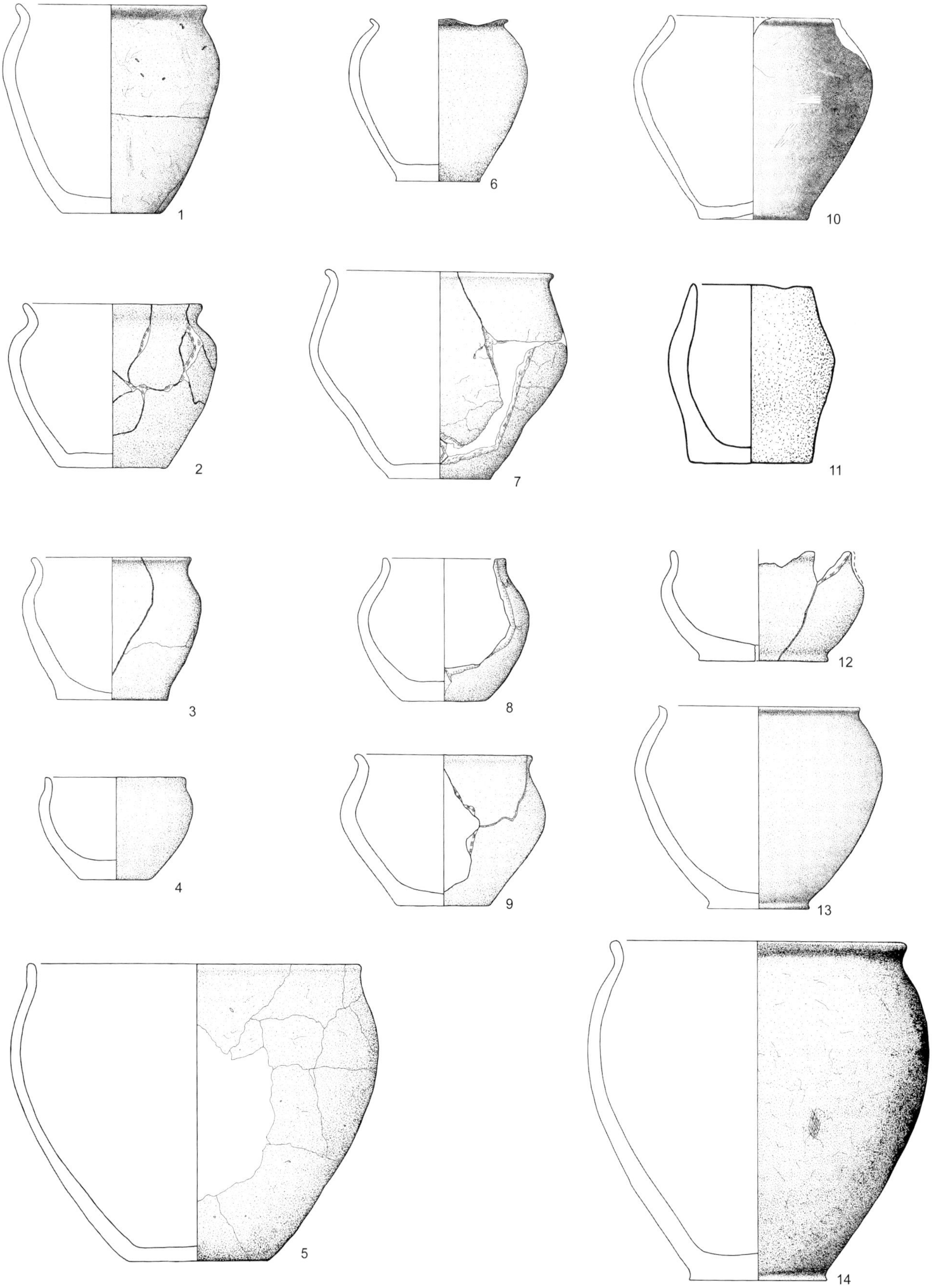

Plate 4. *Groß Strömkendorf, Lkr. Nordwestmecklenburg. Sukow pottery from the trading site. 1–10, 12–14 scale 1:4; 11 scale 1:2.*

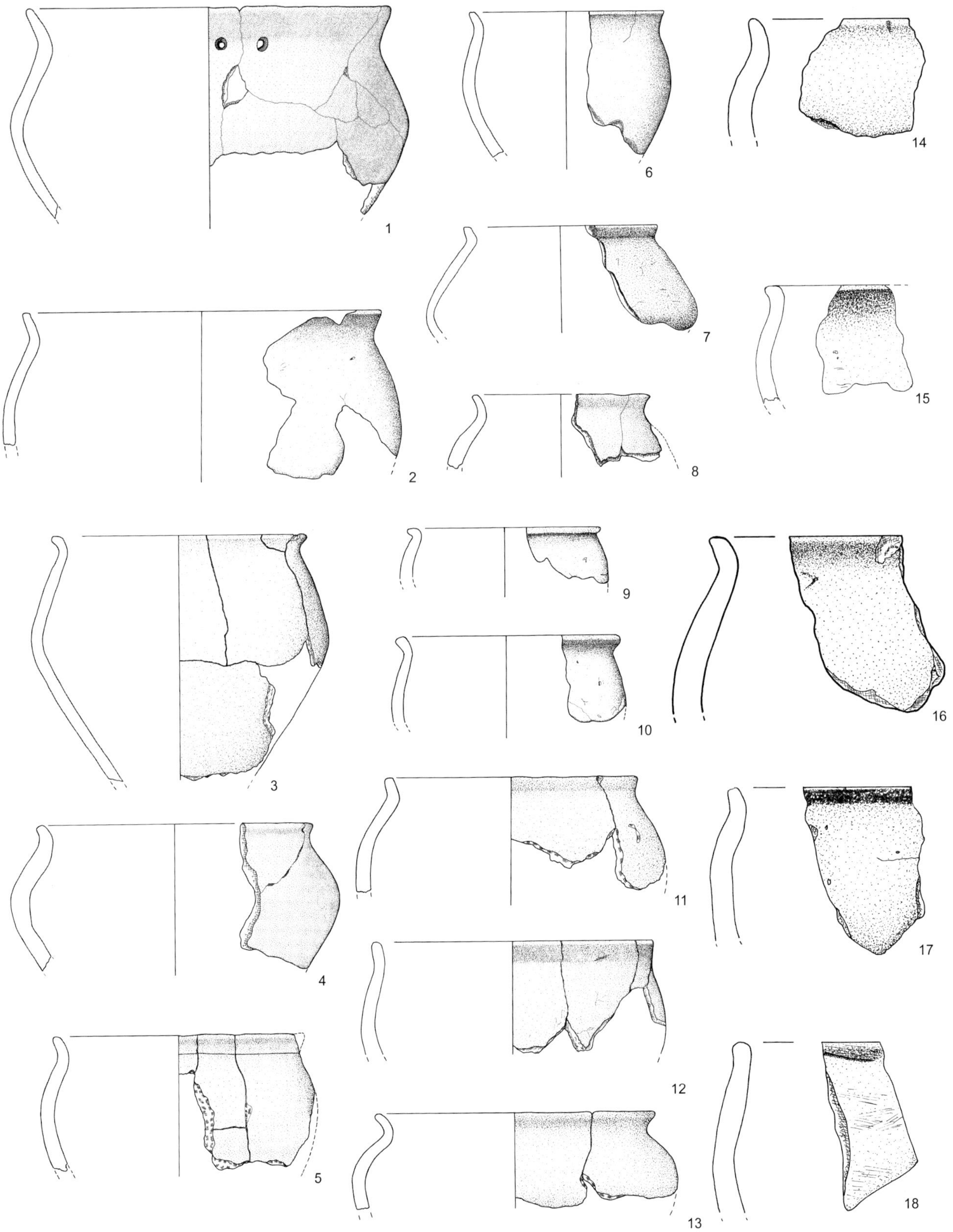

Plate 5. *Groß Strömkendorf, Lkr. Nordwestmecklenburg. Sukow pottery from the trading site. 1–13 scale 1:4; 14–18 scale 1:2.*

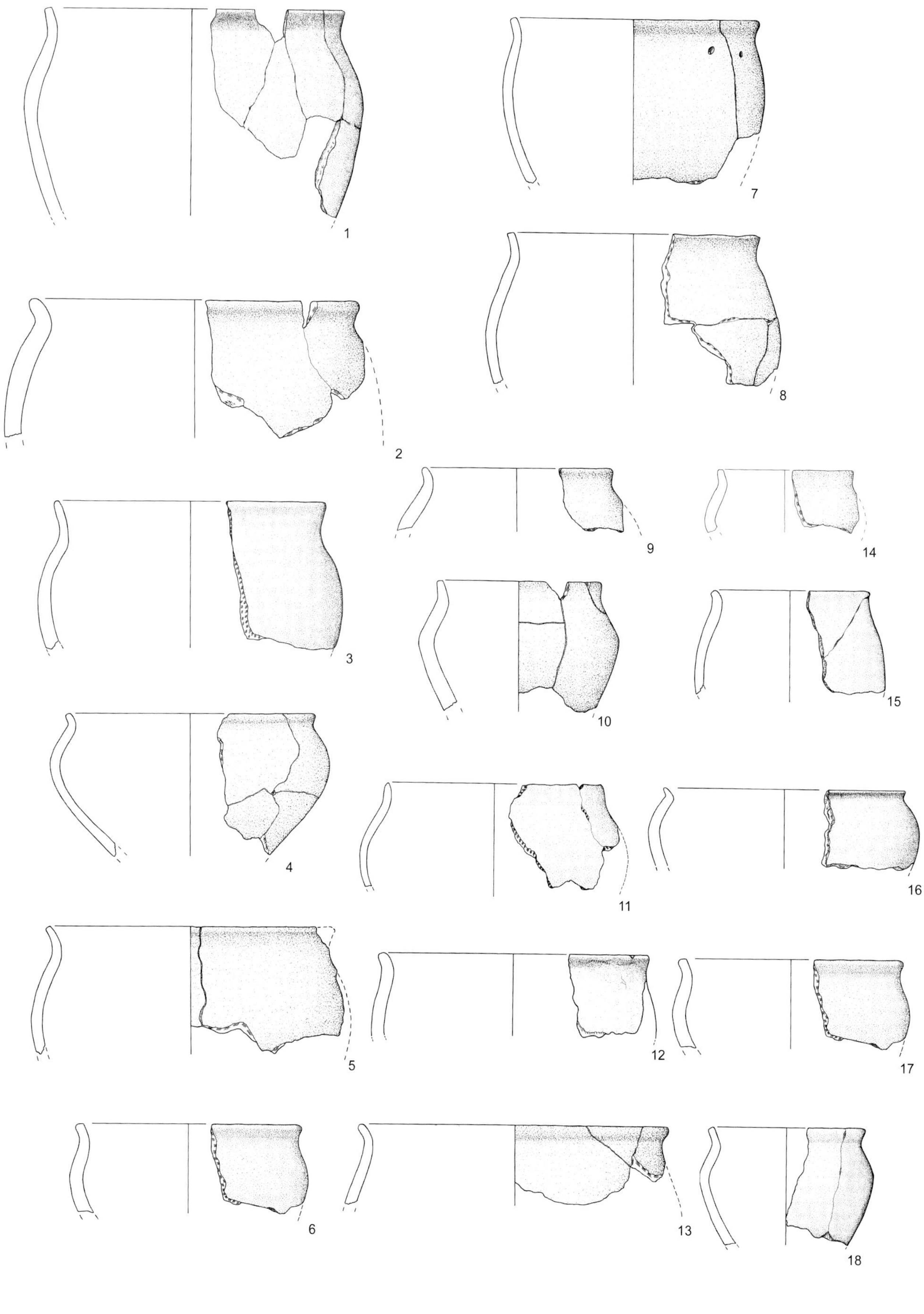

Plate 6. *Groß Strömkendorf, Lkr. Nordwestmecklenburg. Sukow pottery from the trading site. Scale 1:4.*

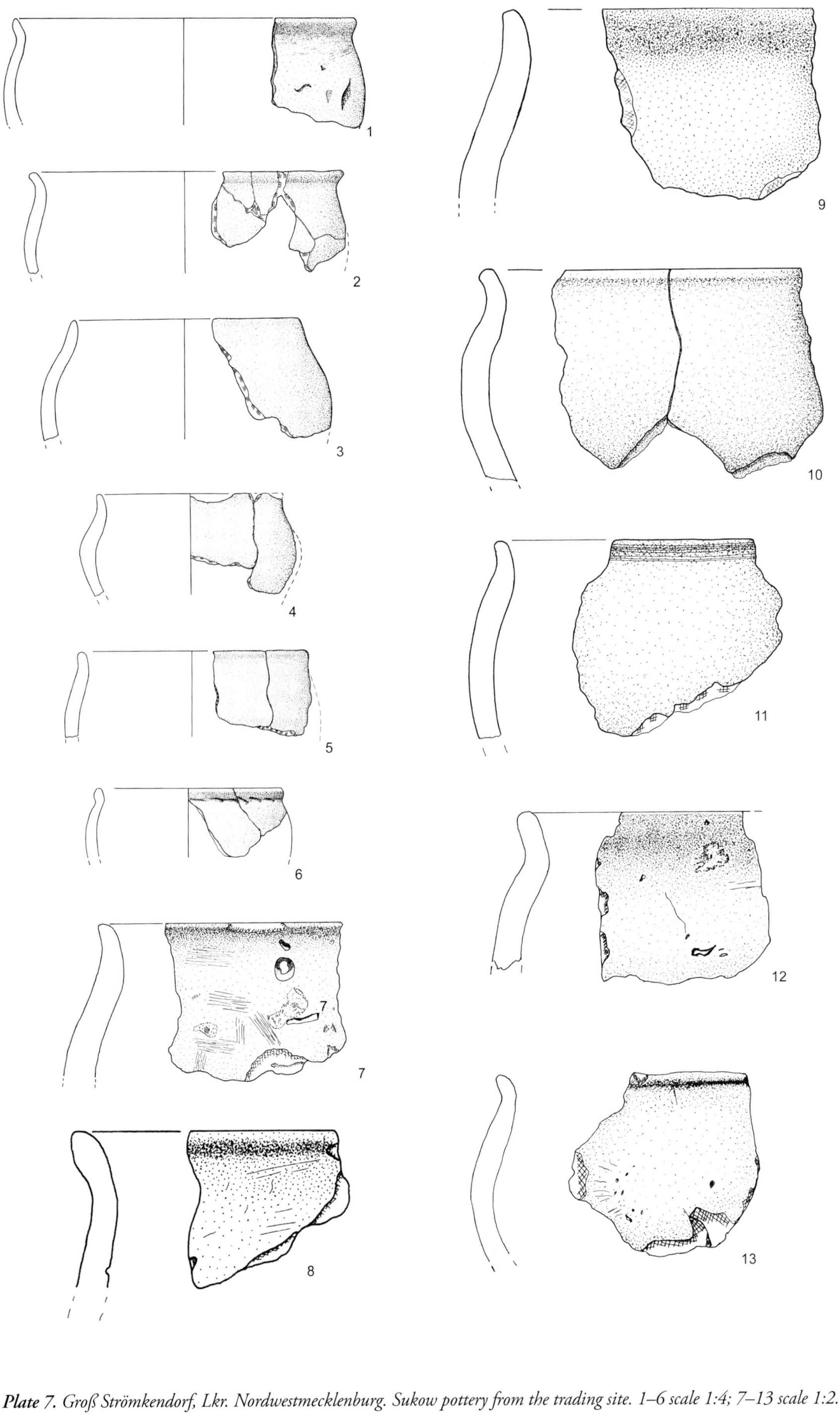

Plate 7. *Groß Strömkendorf, Lkr. Nordwestmecklenburg. Sukow pottery from the trading site. 1–6 scale 1:4; 7–13 scale 1:2.*

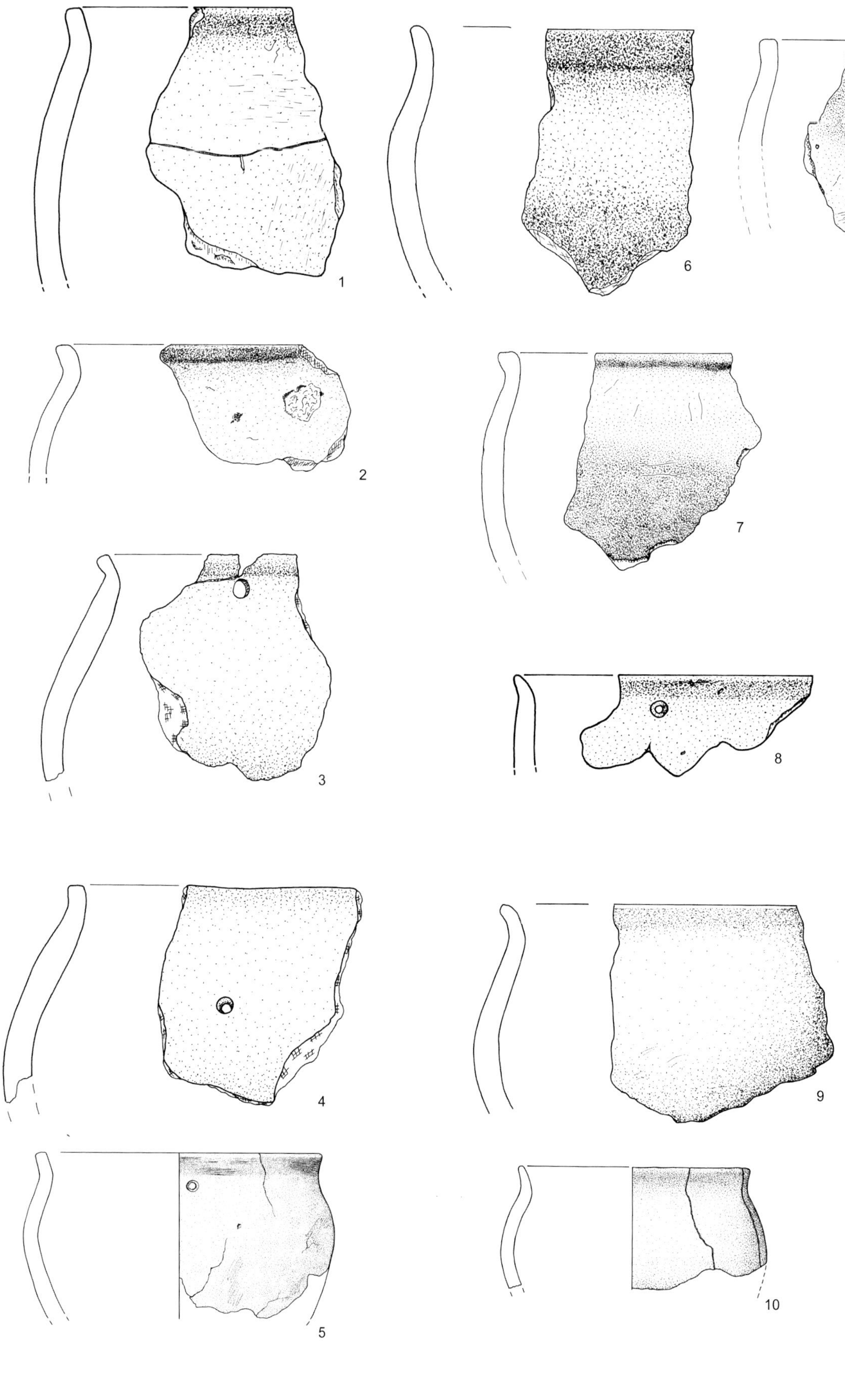

Plate 8. *Groß Strömkendorf, Lkr. Nordwestmecklenburg. Sukow pottery from the trading site. 1–4, 6–9, 11 scale 1:2; 5, 10 scale 1:4.*

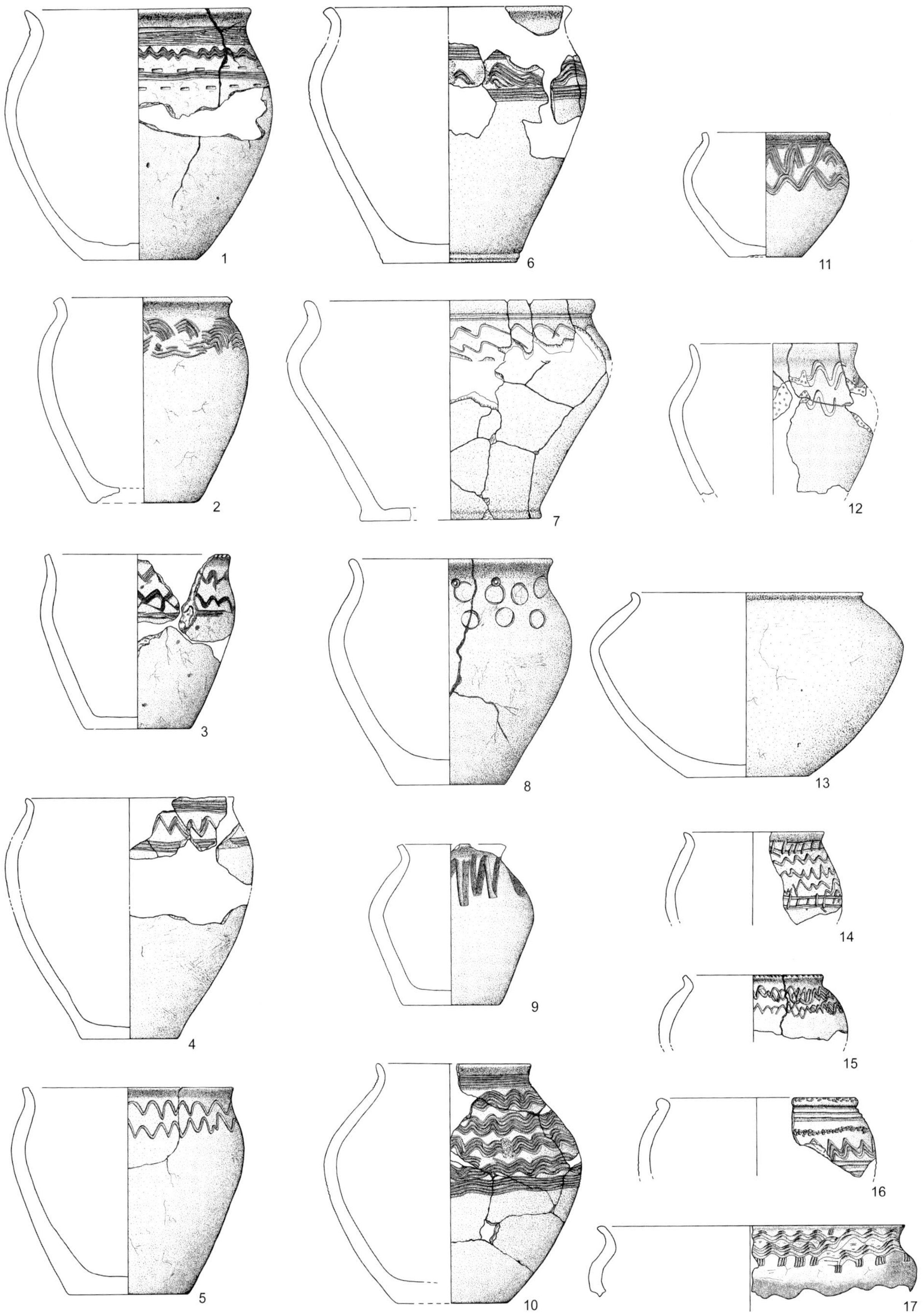

Plate 9. *Groß Strömkendorf, Lkr. Nordwestmecklenburg. Feldberg pottery from the cemetery. Scale 1:4.*

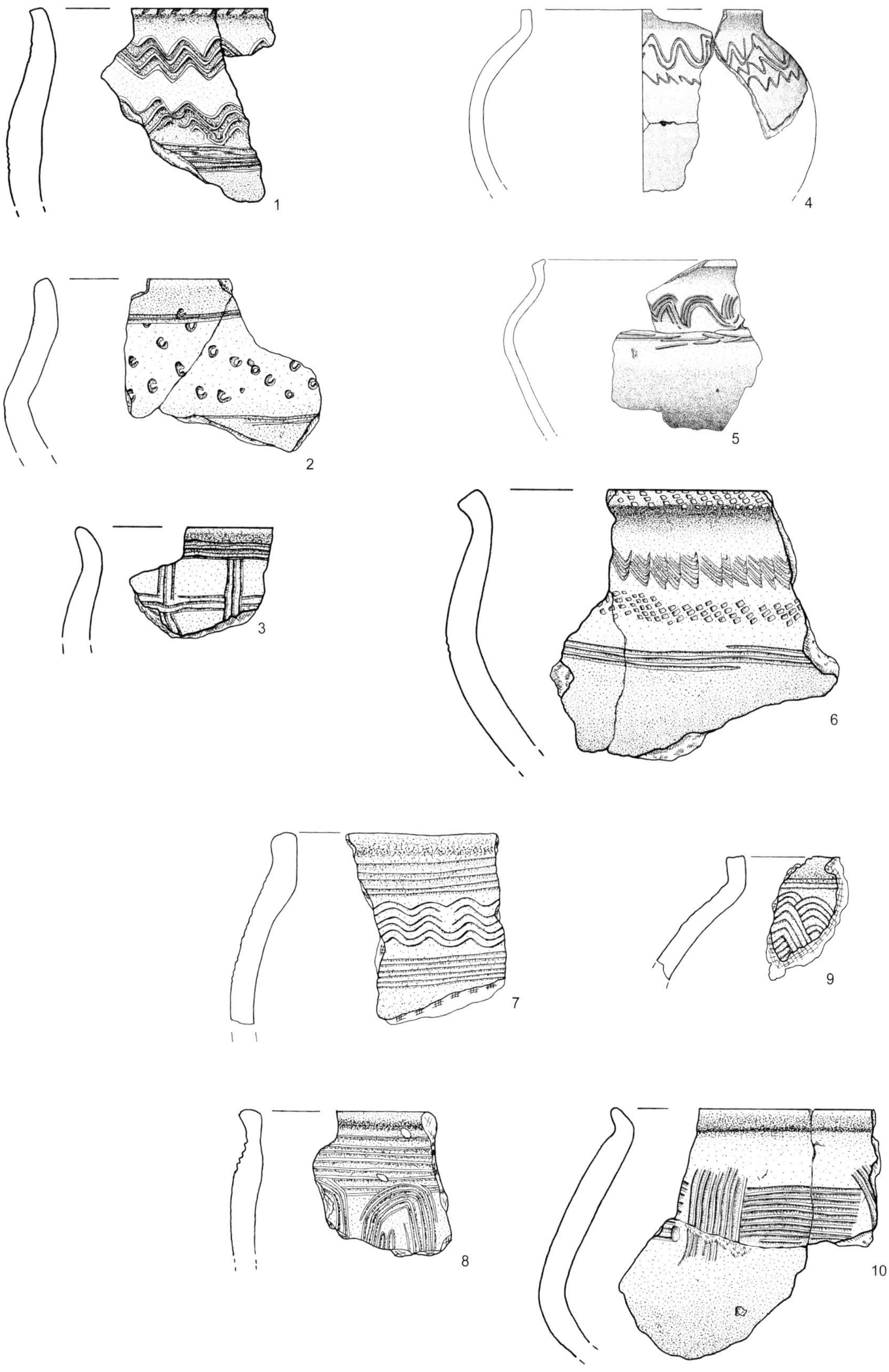

__Plate 10.__ Groß Strömkendorf, Lkr. Nordwestmecklenburg. 1–3 Feldberg pottery from the cemetery; 4–10 Feldberg pottery from the trading site. 1–3, 6–10 scale 1:2; 4–5 scale 1:4.

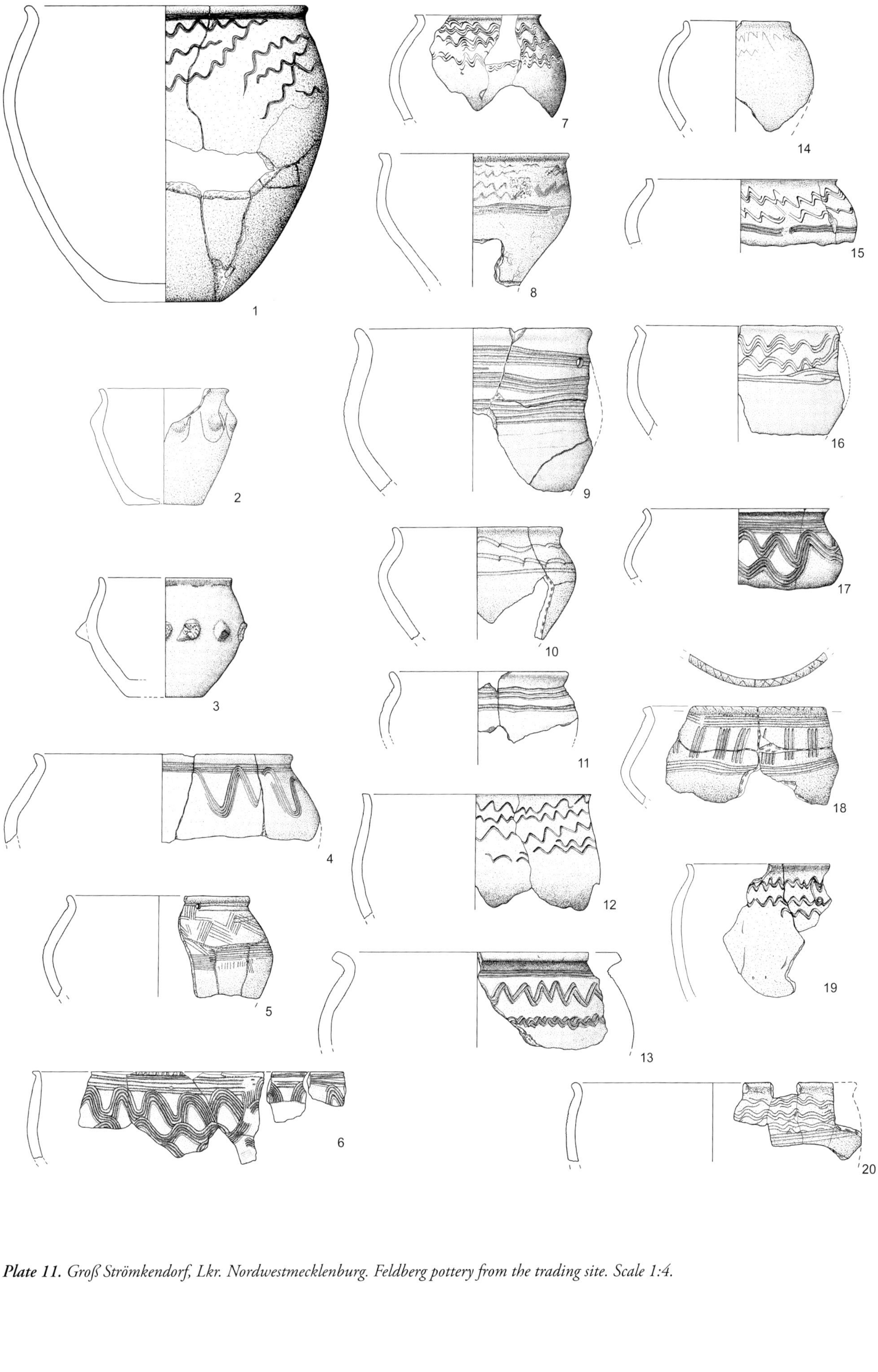

Plate 11. *Groß Strömkendorf, Lkr. Nordwestmecklenburg. Feldberg pottery from the trading site. Scale 1:4.*

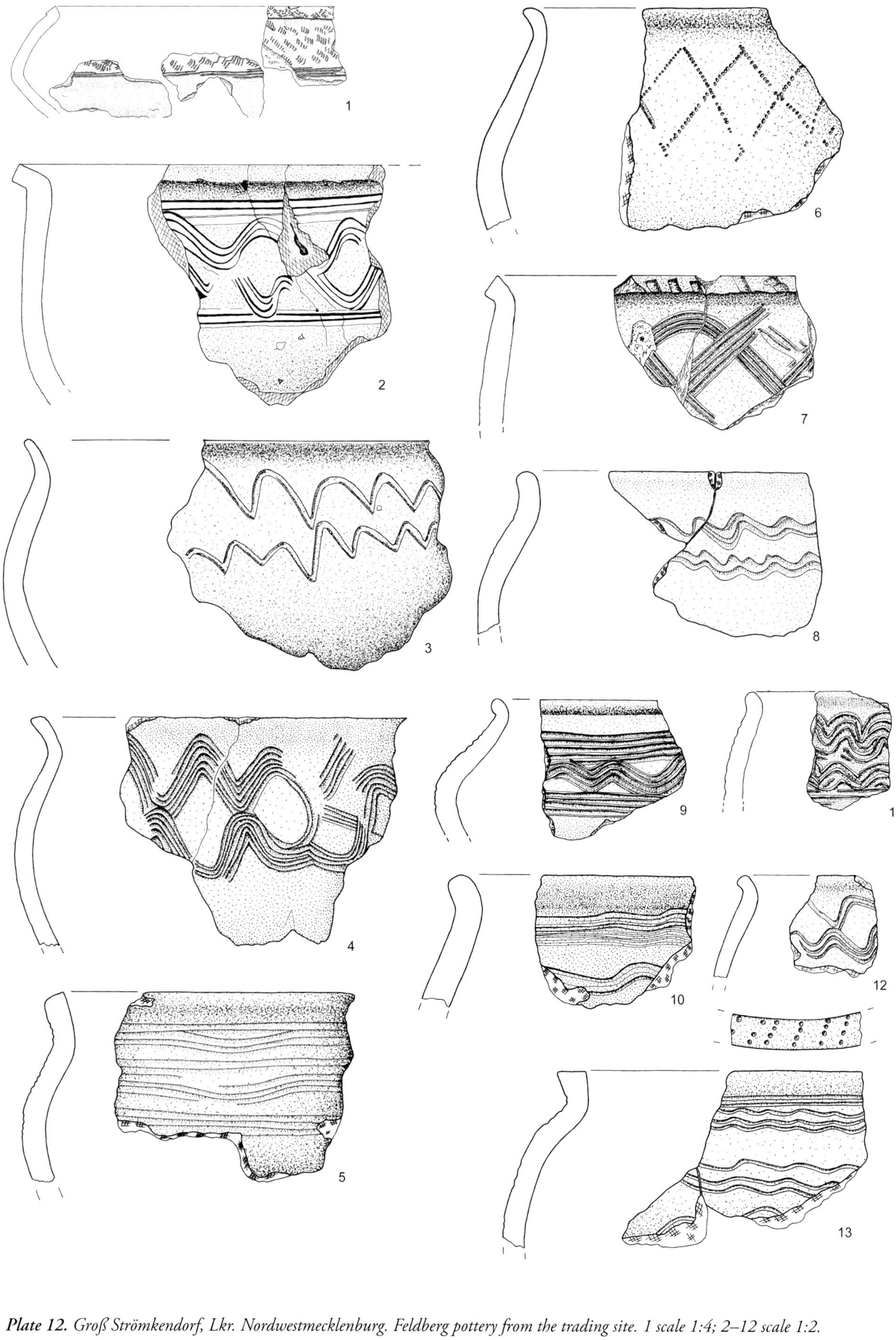

Plate 12. *Groß Strömkendorf, Lkr. Nordwestmecklenburg. Feldberg pottery from the trading site. 1 scale 1:4; 2–12 scale 1:2.*

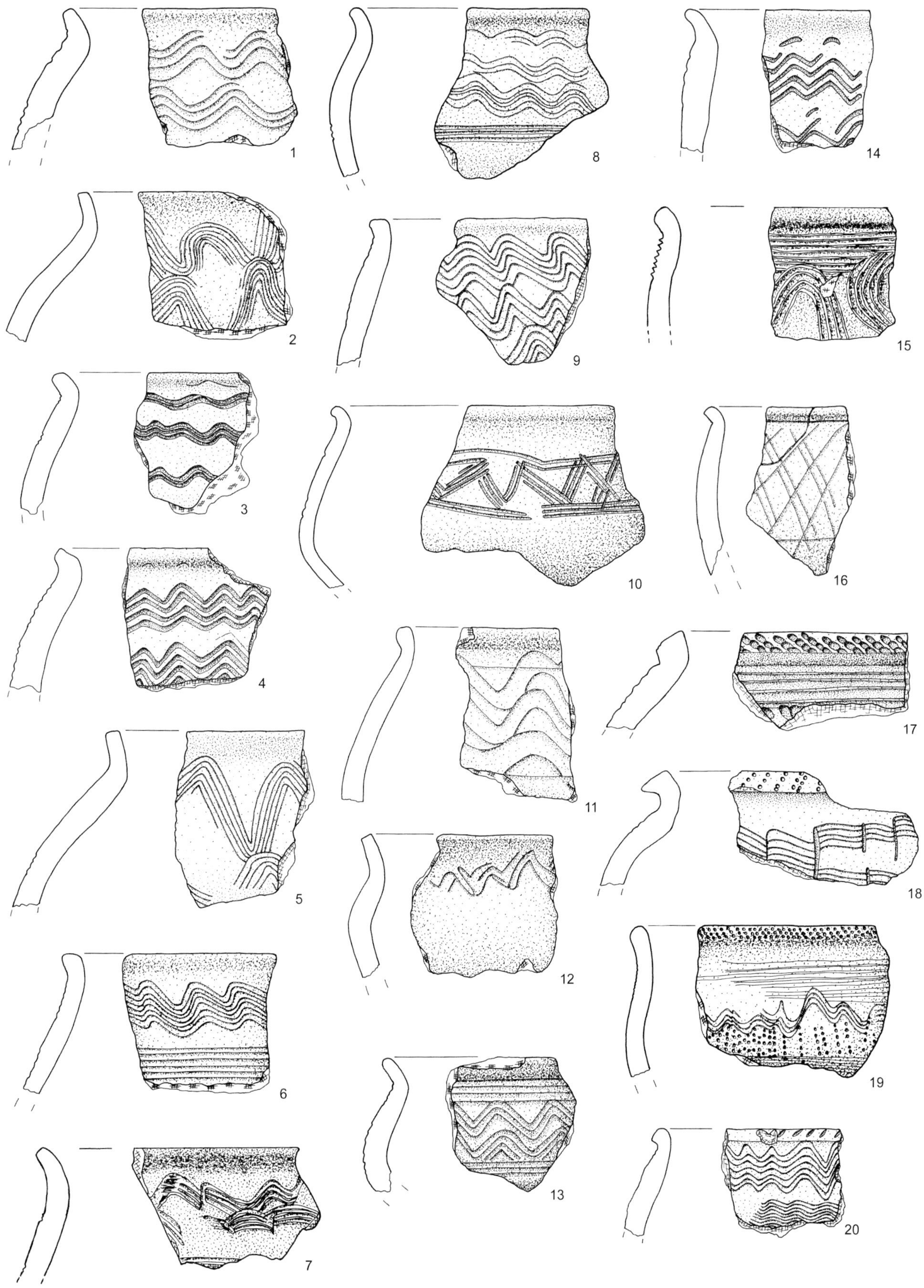

Plate 13. *Groß Strömkendorf, Lkr. Nordwestmecklenburg. Feldberg pottery from the trading site. Scale 1:2.*

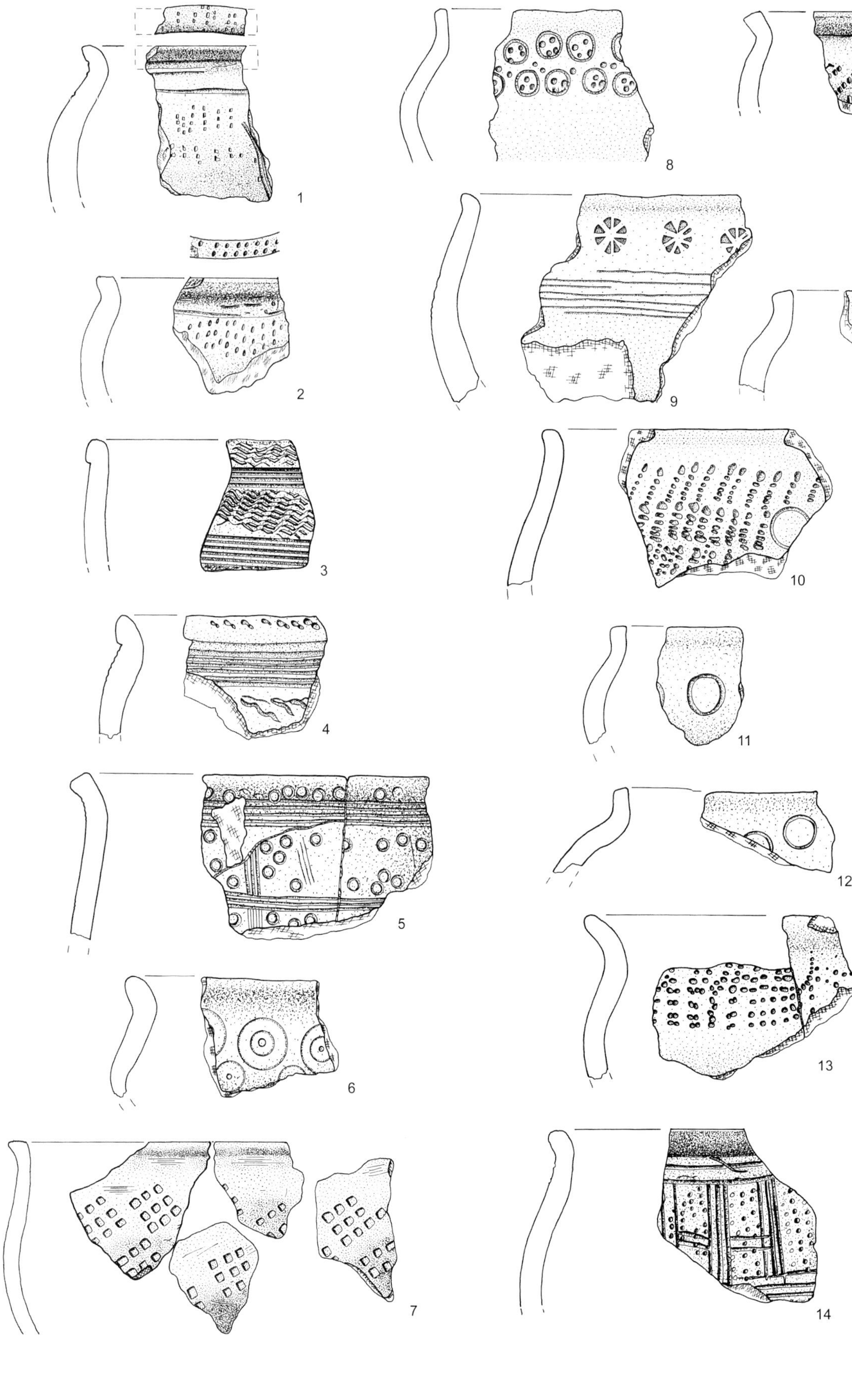

Plate 14. Groß Strömkendorf, Lkr. Nordwestmecklenburg. Feldberg pottery from the trading site. Scale 1:2.

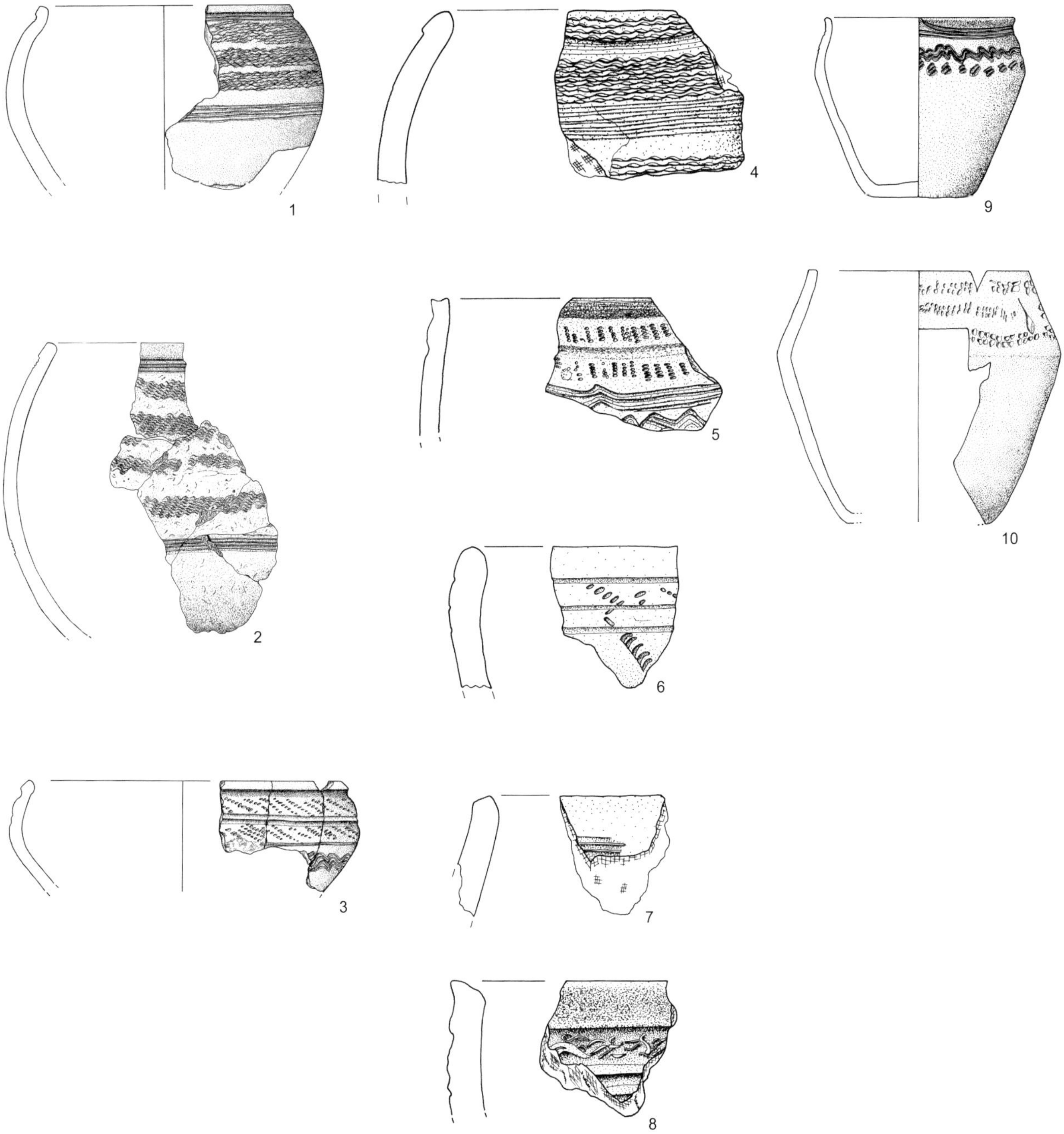

Plate 15. *Groß Strömkendorf, Lkr. Nordwestmecklenburg. 1–7 Fresendorf pottery from the trading site; 8 Fresendorf pottery from the cemetery; 9 Menkendorf pottery from the trading site; 10 Menkendorf pottery from the cemetery. 1–3, 9–10 scale 1:4; 4–8 scale 1:2.*

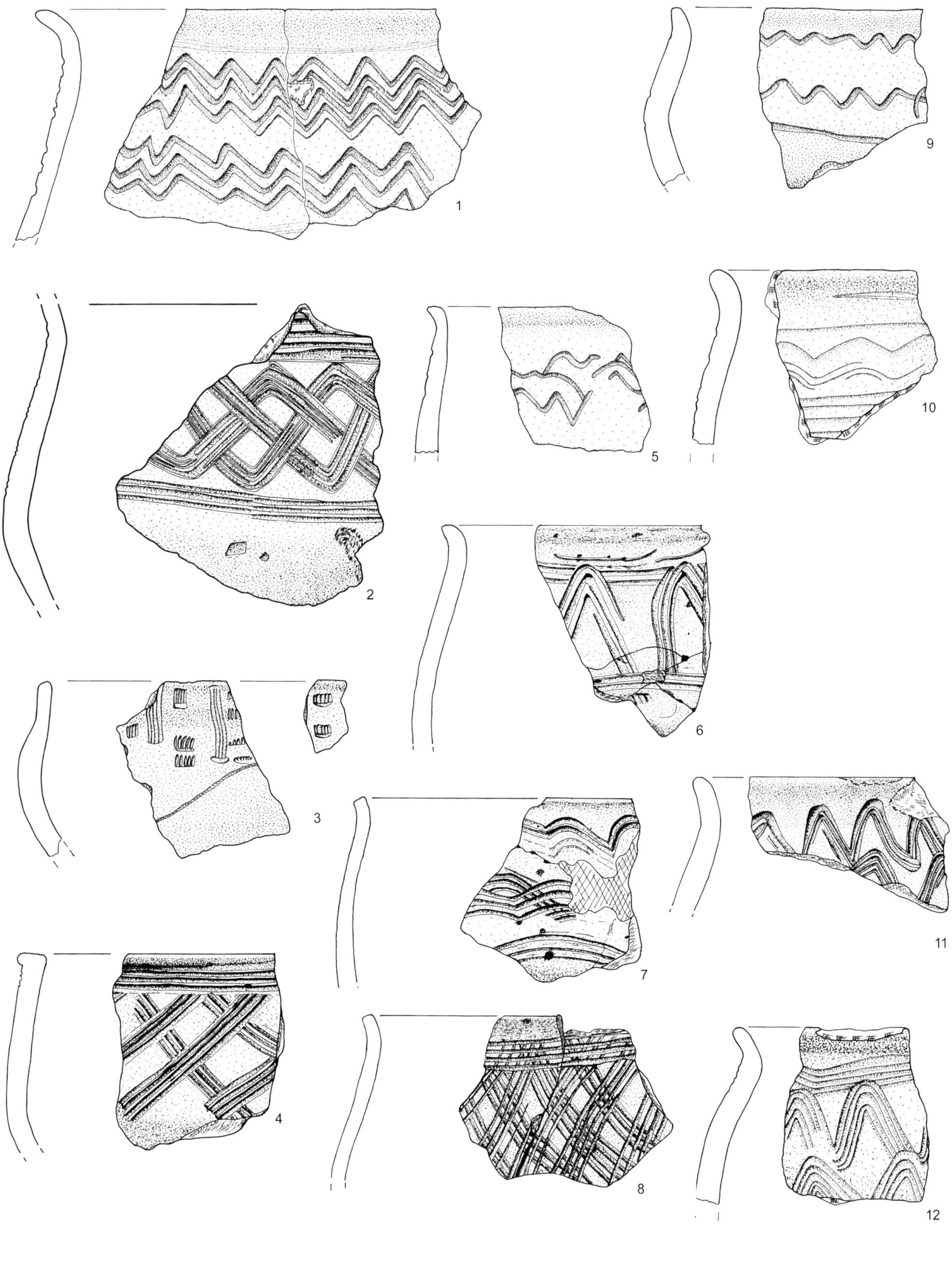

Plate 16. *Groß Strömkendorf, Lkr. Nordwestmecklenburg. 1, 4–12 Menkendorf pottery from the trading site; 2–3 Menkendorf pottery from the cemetery. 1 scale 1:4; 2–12 scale 1:2.*

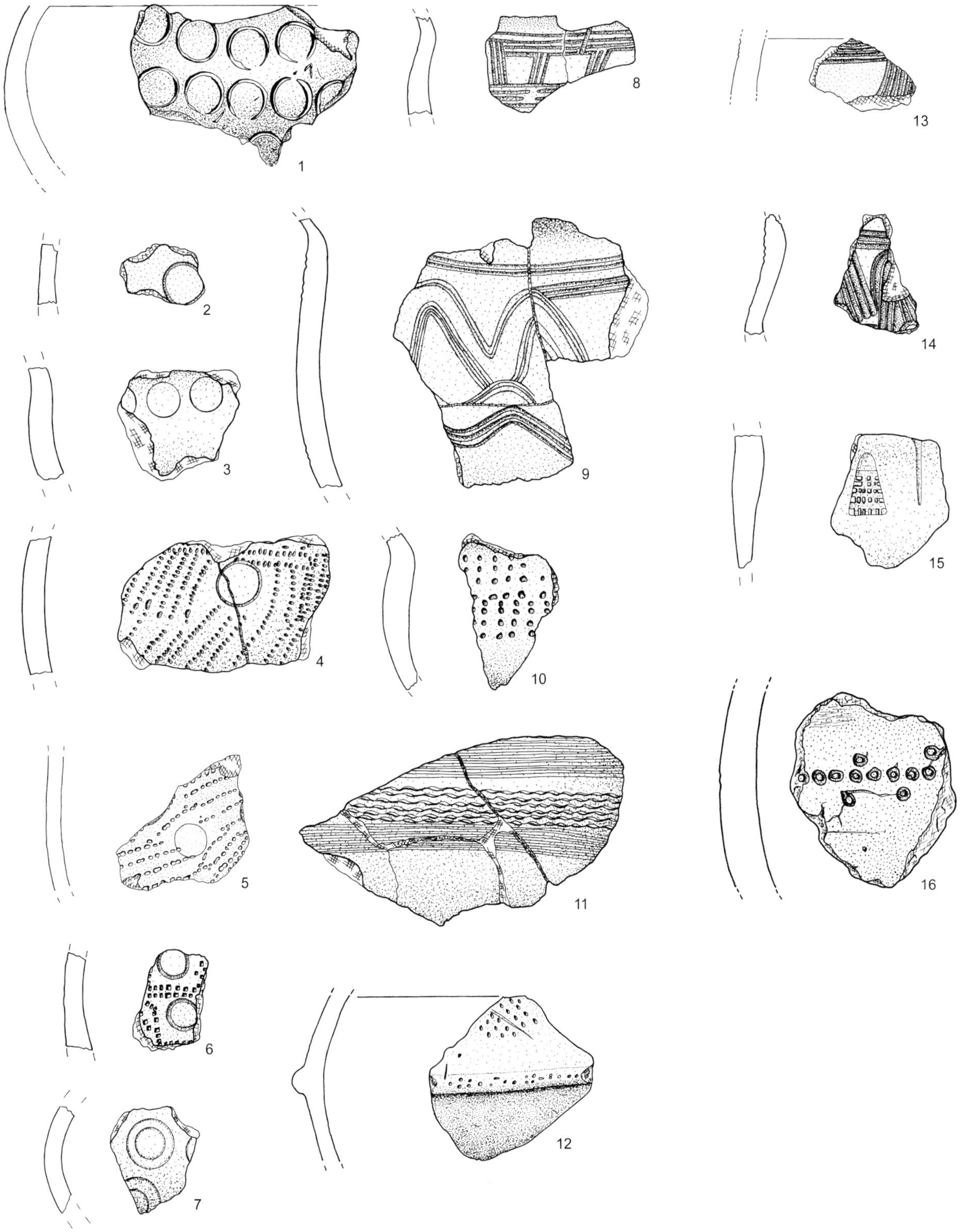

Plate 17. *Groß Strömkendorf, Lkr. Nordwestmecklenburg. 1–15 Slavonic pottery from the trading site; 16 Slavonic pottery from the cemetery. Scale 1:2.*

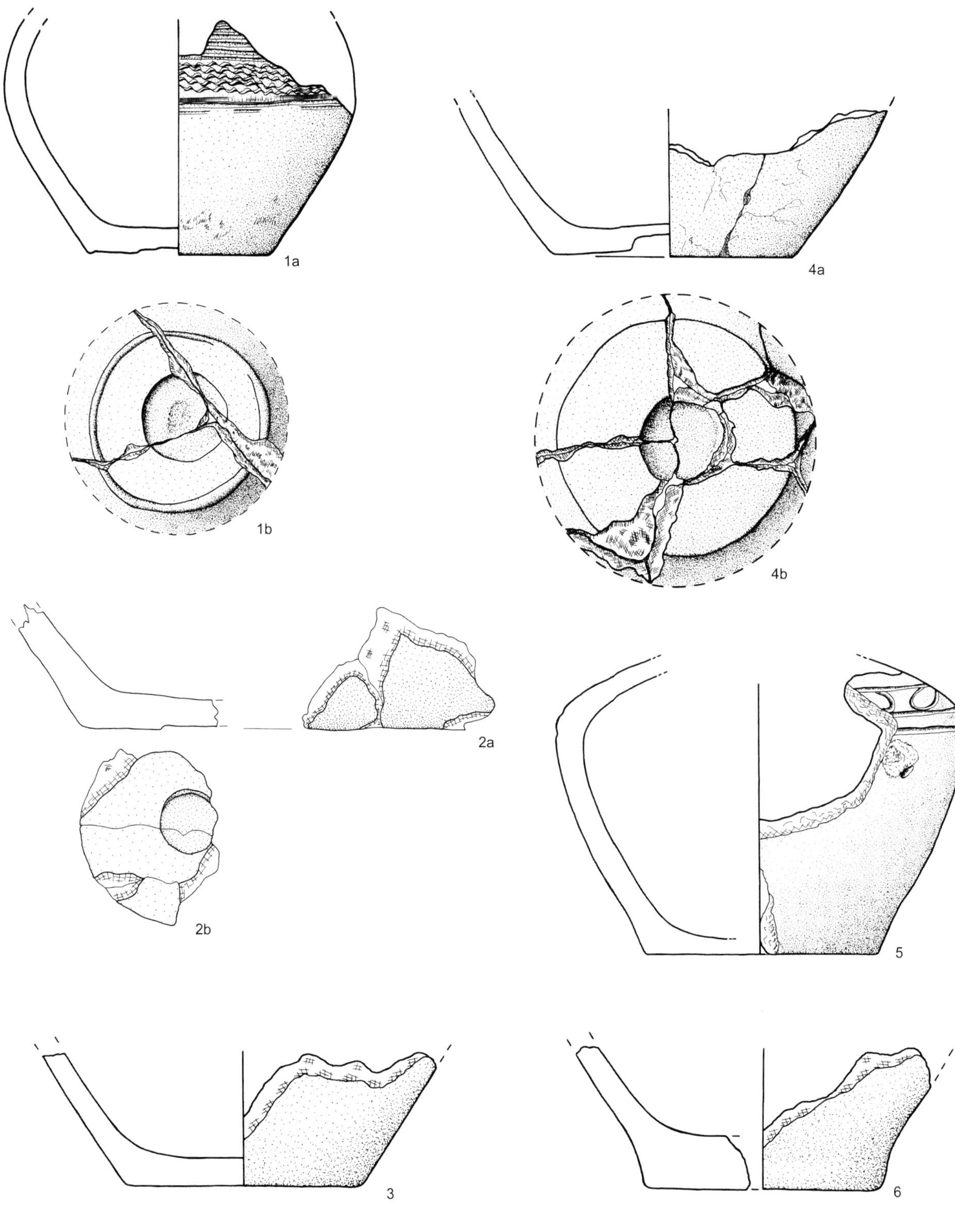

Plate 18. *Groß Strömkendorf, Lkr. Nordwestmecklenburg. 1 Base sherd from the cemetery; 2–6 Base sherds from the trading site. Scale 1:2.*

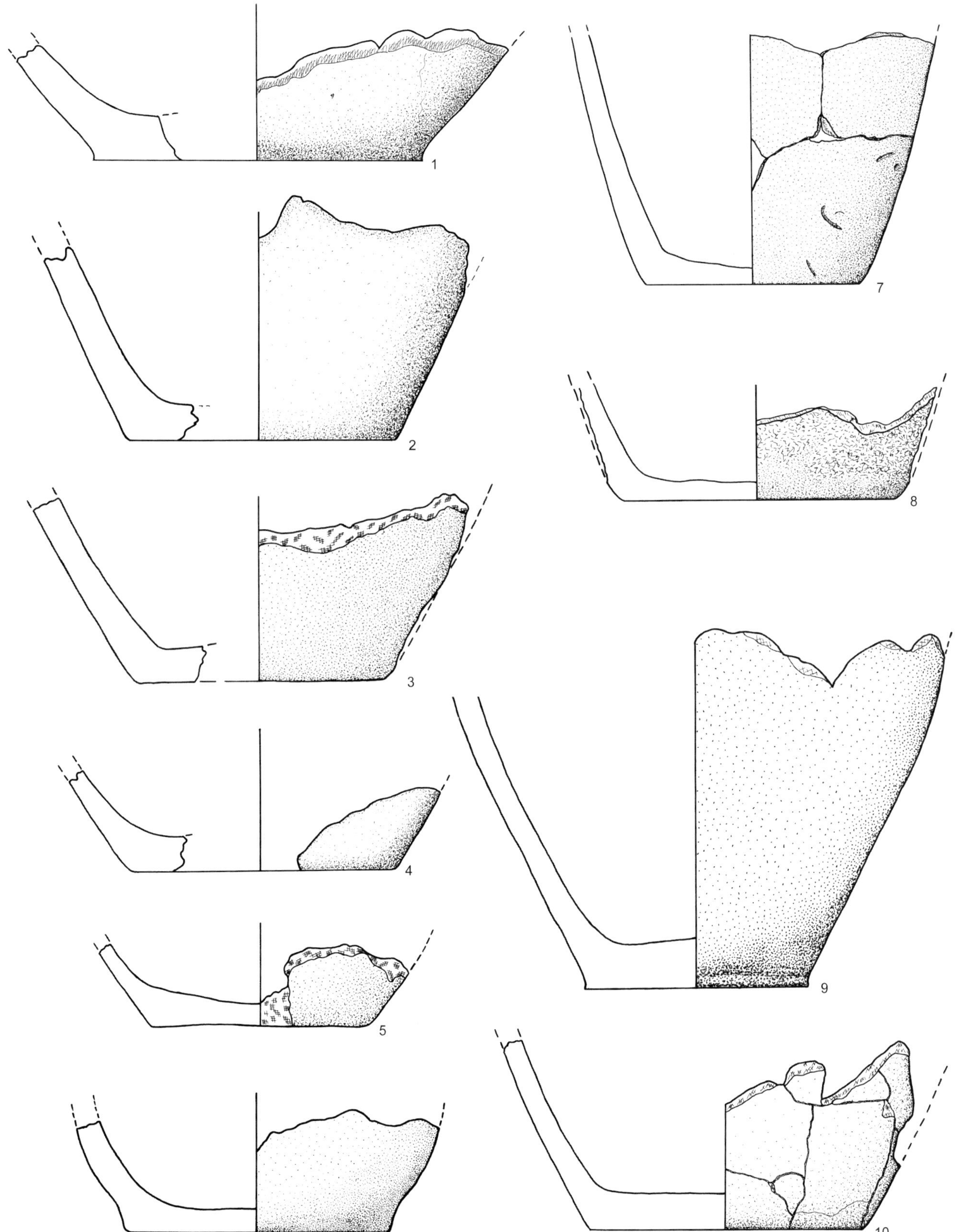

Plate 19. *Groß Strömkendorf, Lkr. Nordwestmecklenburg. 1–5 Base sherds from the trading site; 6–10 Base sherds from the cemetery. Scale 1:2.*

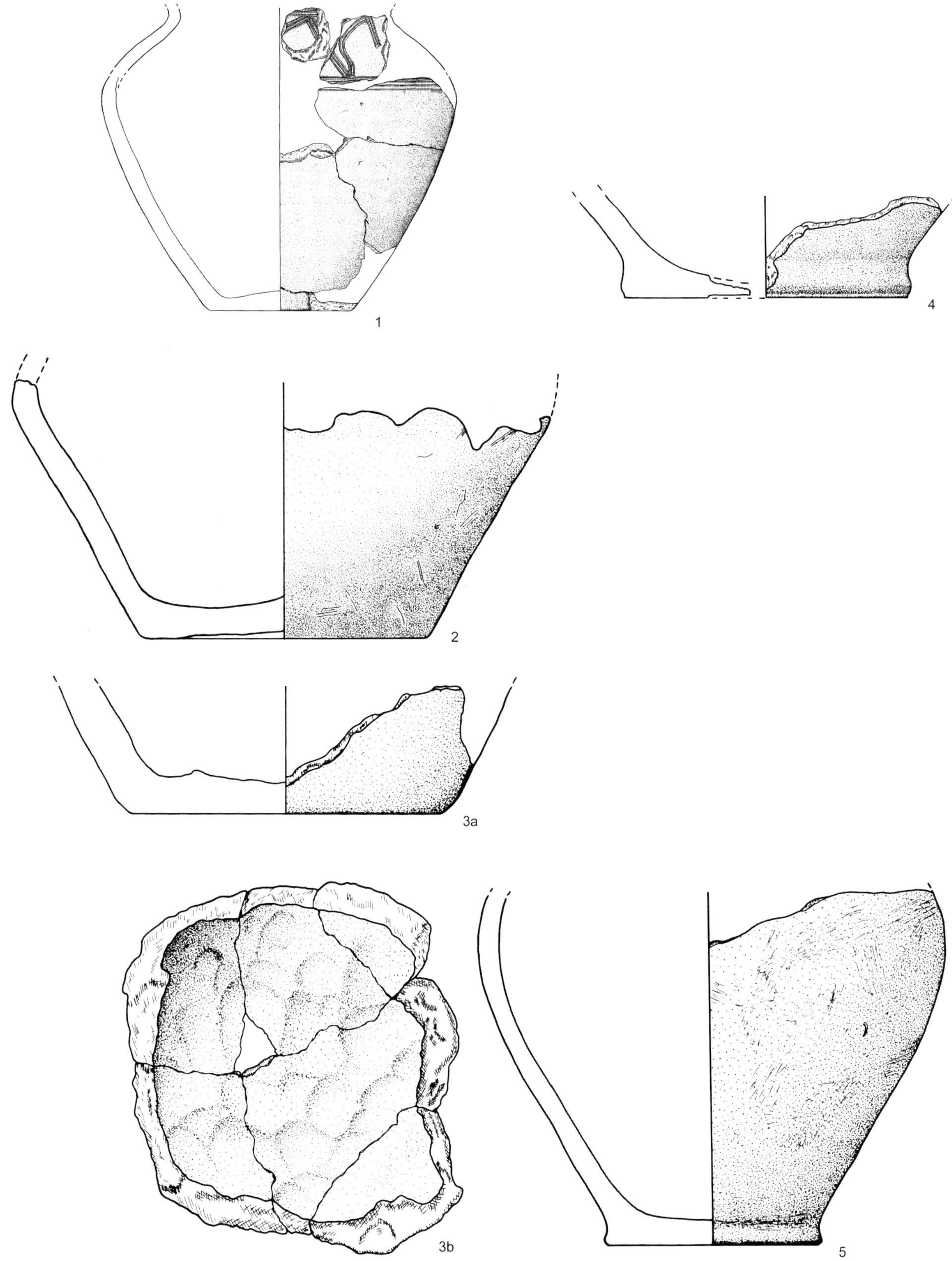

***Plate 20.** Groß Strömkendorf, Lkr. Nordwestmecklenburg. Base sherds from the cemetery. 1 scale 1:4; 2–5 scale 1:2.*

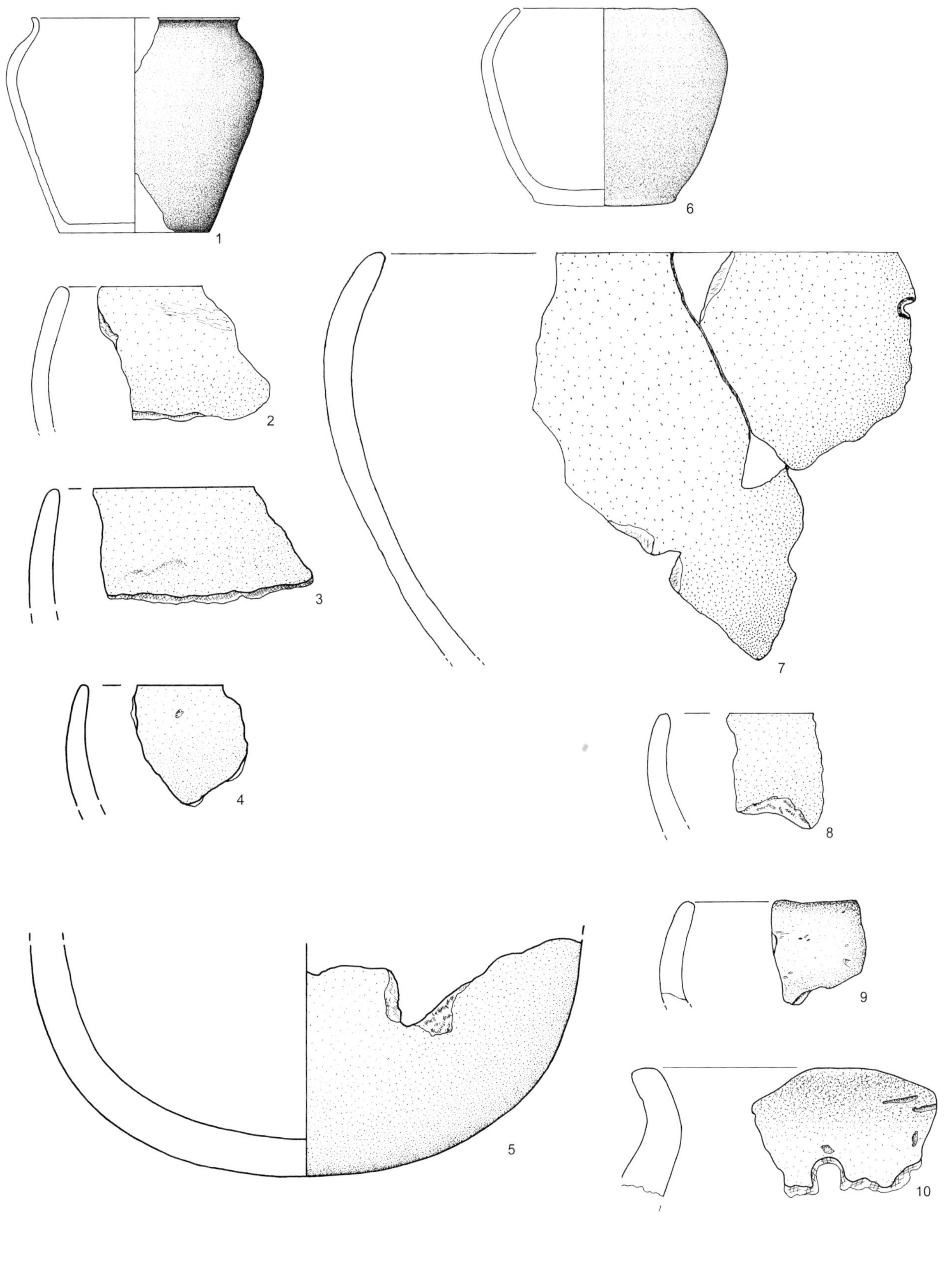

Plate 21. *Groß Strömkendorf, Lkr. Nordwestmecklenburg. 1–5 Scandinavian pottery from the cemetery; 6–10 Scandinavian pottery from the trading site. 1, 6 scale 1:4; 2–5, 7–10 scale 1:2.*

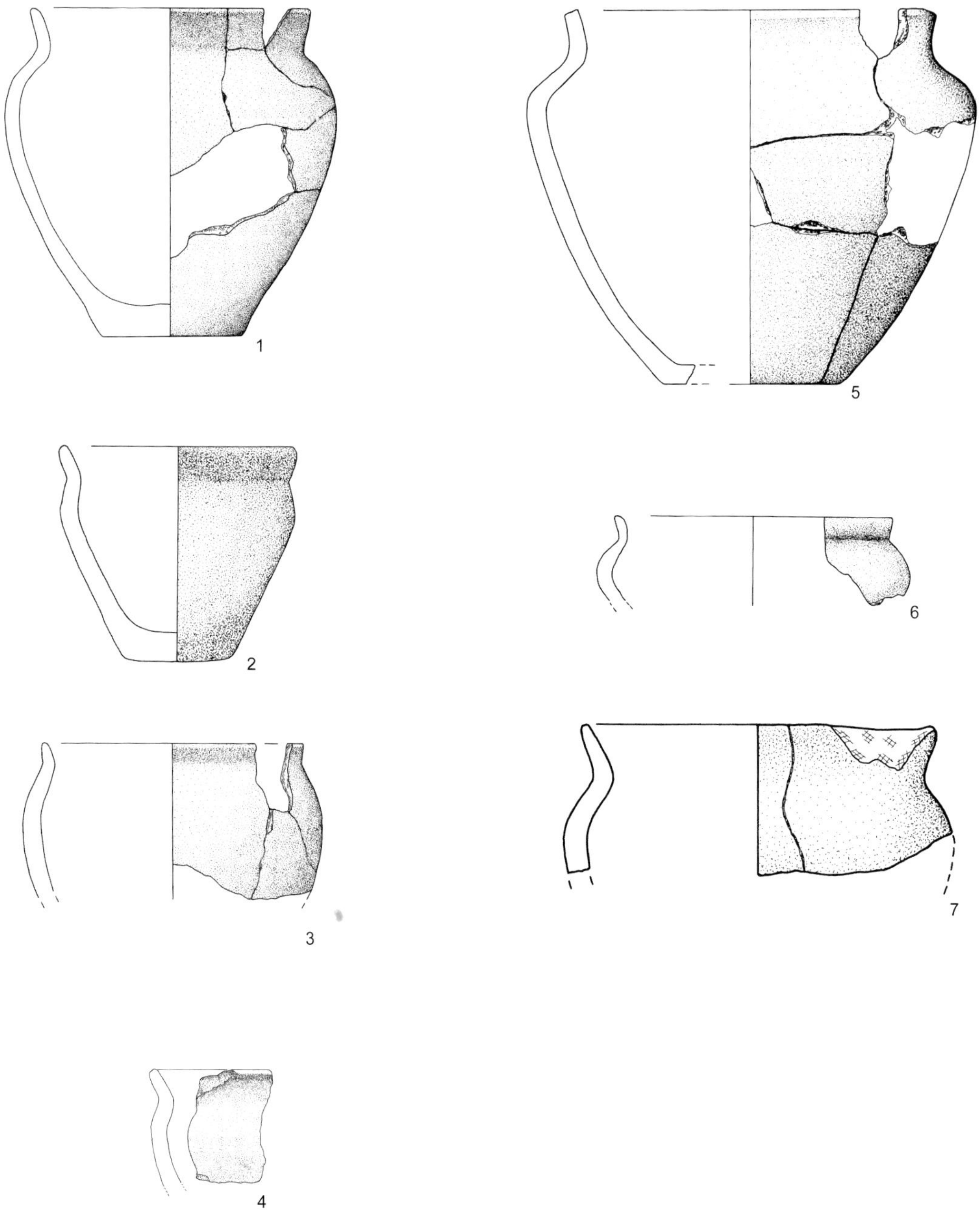

Plate 22. *Groß Strömkendorf, Lkr. Nordwestmecklenburg. 1–4 Saxon pottery from the cemetery; 5–7 Saxon pottery from the trading site. 1–3, 5–7 scale 1:4; 4 scale 1:2.*

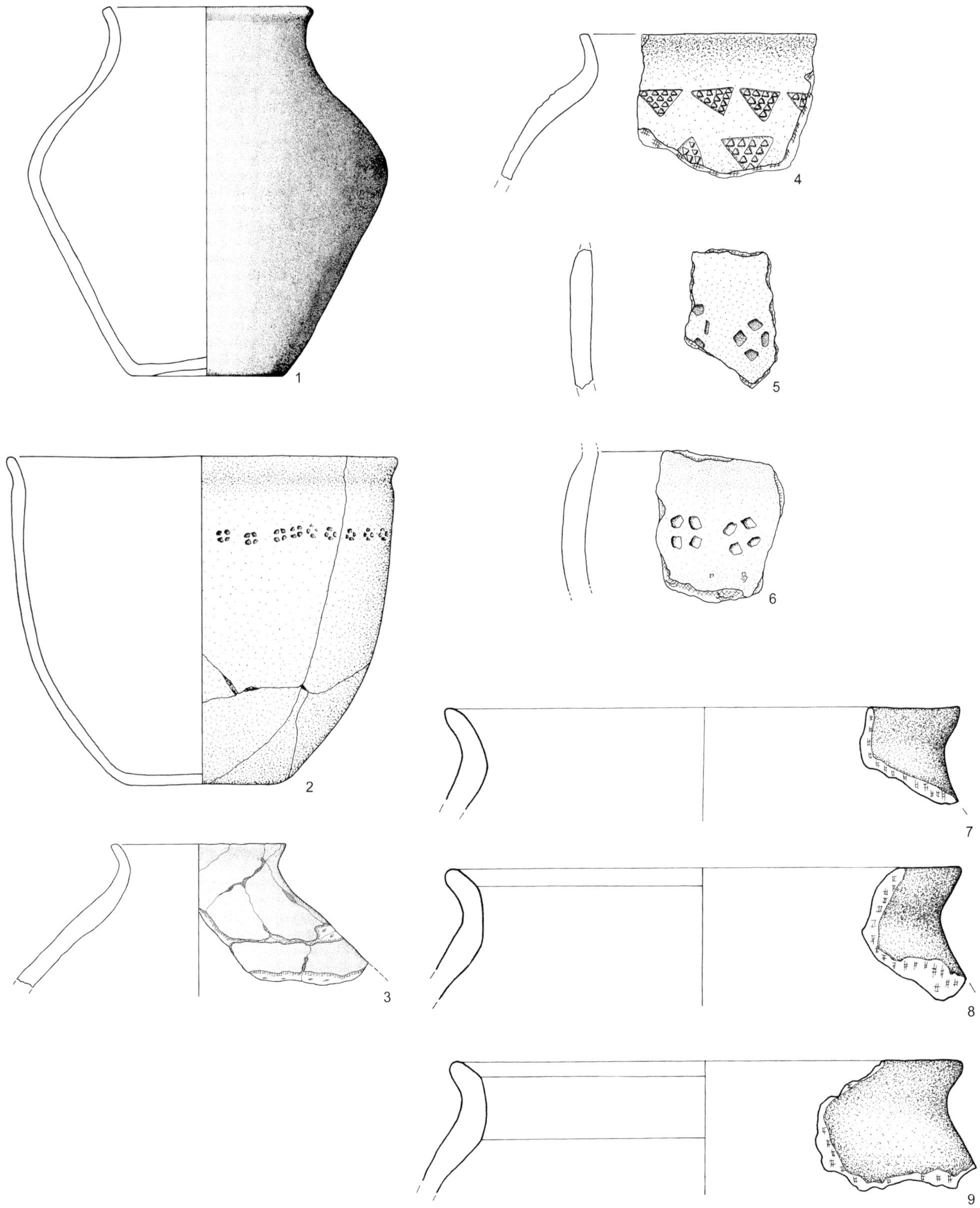

Plate 23. *Groß Strömkendorf, Lkr. Nordwestmecklenburg. 1 Frisian pottery from the cemetery; 2–6 Frisian pottery from the trading site; 7–9 Muschelgrus ware from the trading site. 1–3 scale 1:4; 4–9 scale 1:2.*

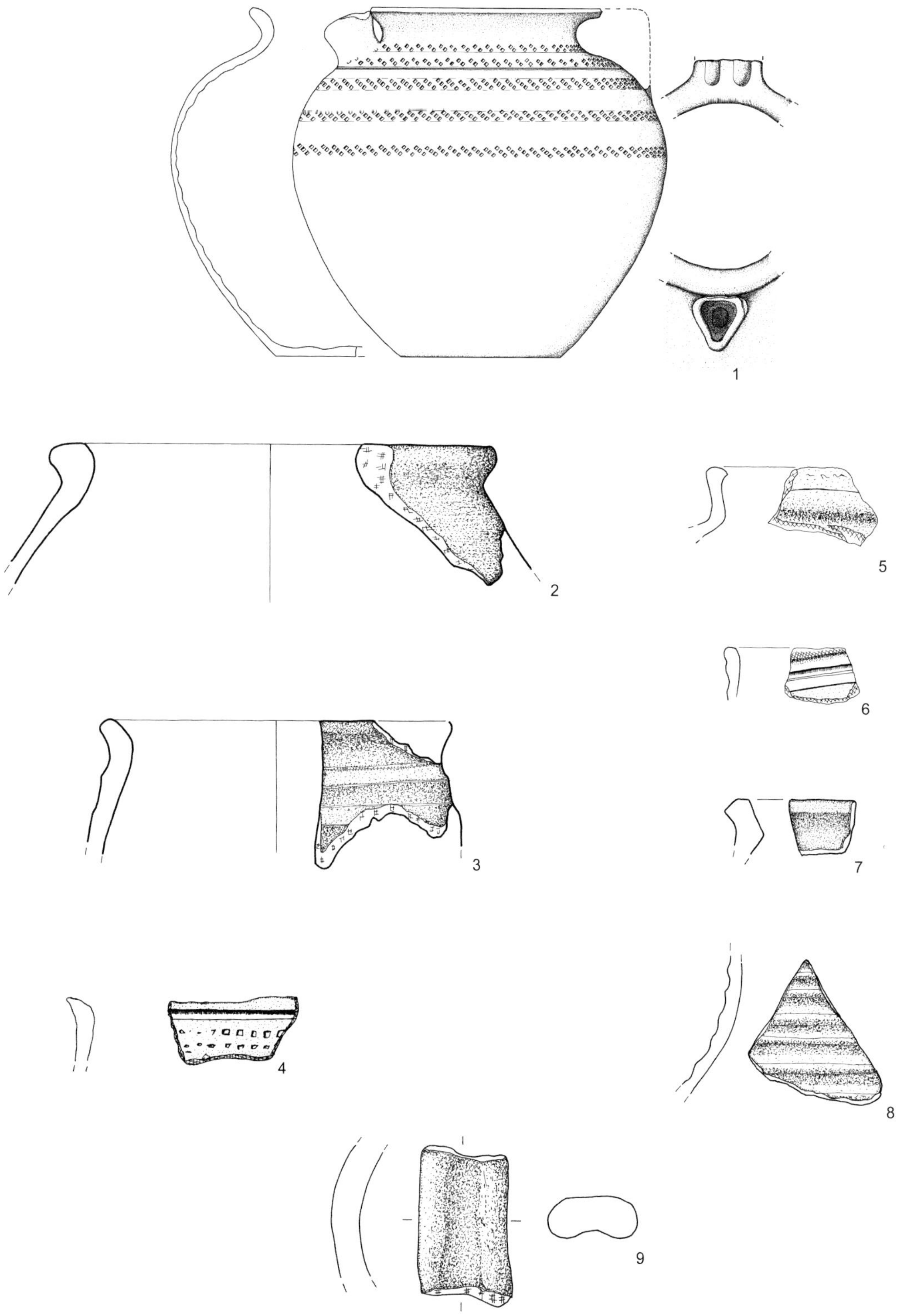

Plate 24. *Groß Strömkendorf, Lkr. Nordwestmecklenburg. 1 Pottery of Merovingian Black Ware from the cemetery; 2–9 Pottery of Merovingian Black Ware from the trading site. 1 scale 1:4; 2–9 scale 1:2.*

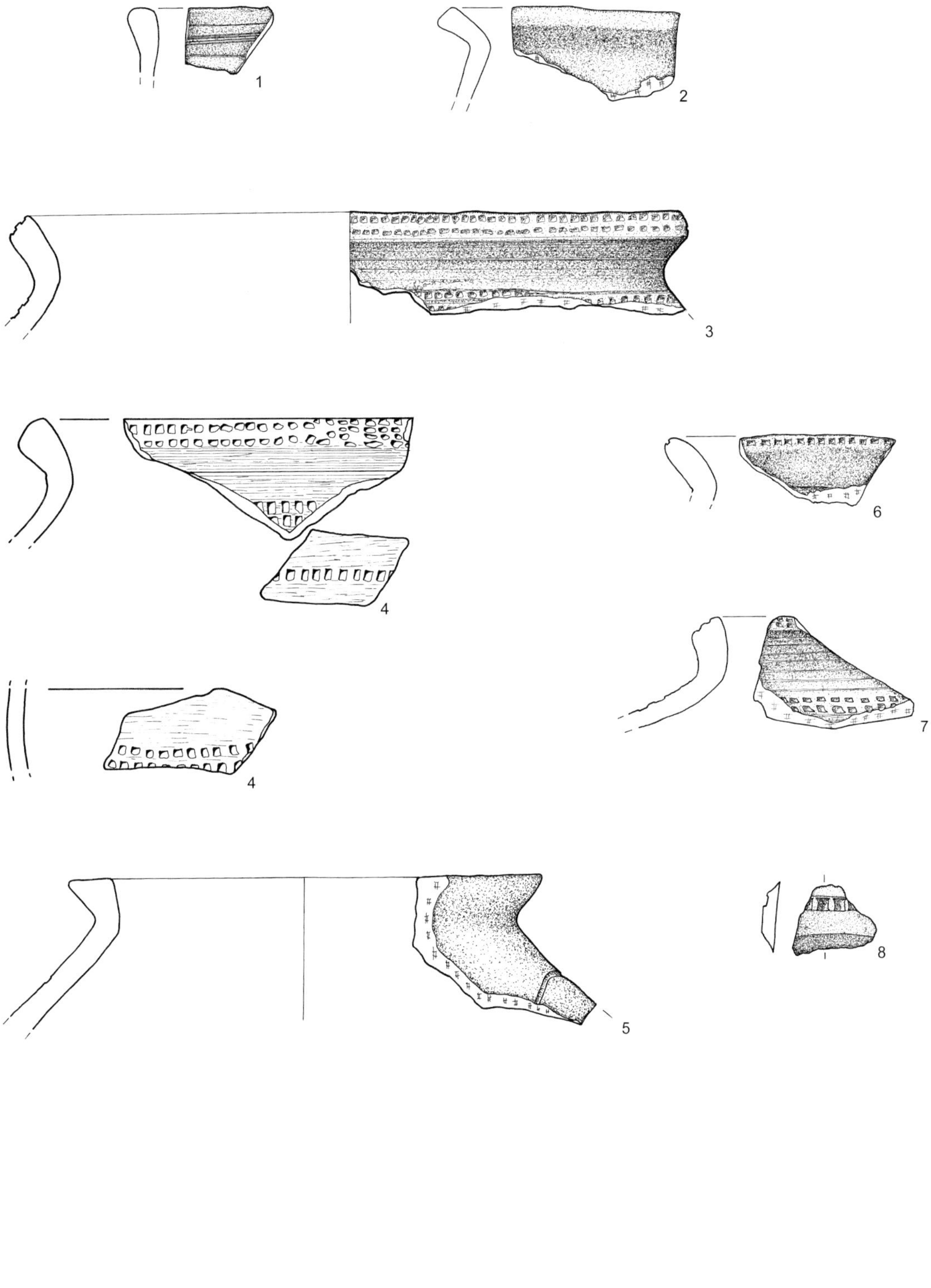

Plate 25. *Groß Strömkendorf, Lkr. Nordwestmecklenburg. 1 Tating pottery from the trading site; 2 Mayen pottery from the trading site; 3–7 Badorf pottery from the trading site; 8 Relief-band amphora (Badorf) from the trading site. Scale 1:2.*

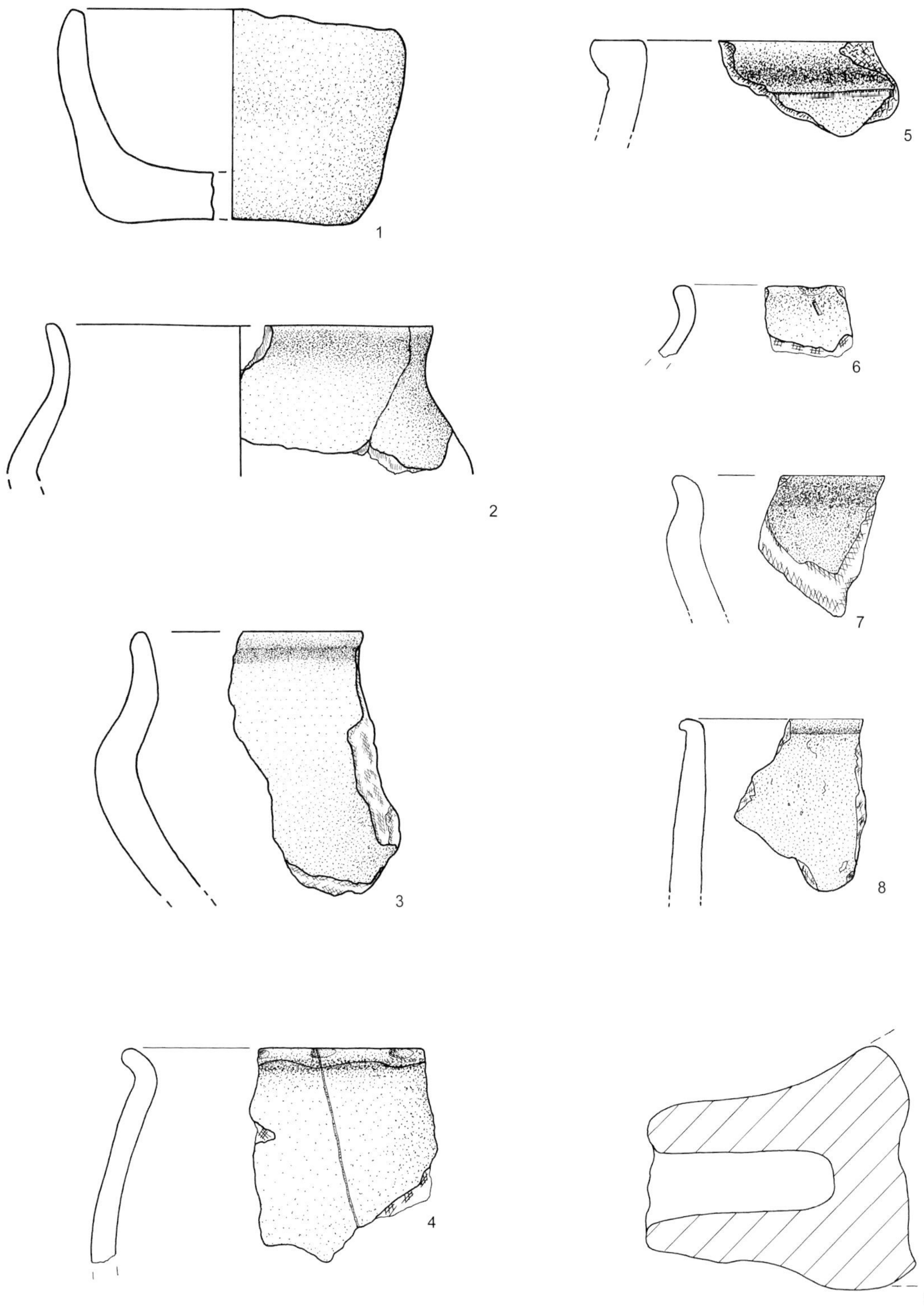

Plate 26. *Groß Strömkendorf, Lkr. Nordwestmecklenburg. Pottery of uncertain origin. 1 scale 1:4; 2–9 scale 1:2.*